Selected Titles in This Series

46 **Stephen Lipscomb,** Symmetric inverse semigroups, 1996

45 **George M. Bergman and Adam O. Hausknecht,** Cogroups and co-rings in categories of associative rings, 1996

44 **J. Amorós, M. Burger, K. Corlette, D. Kotschick, and D. Toledo,** Fundamental groups of compact Kähler manifolds, 1996

43 **James E. Humphreys,** Conjugacy classes in semisimple algebraic groups, 1995

42 **Ralph Freese, Jaroslav Ježek, and J. B. Nation,** Free lattices, 1995

41 **Hal L. Smith,** Monotone dynamical systems: an introduction to the theory of competitive and cooperative systems, 1995

40.2 **Daniel Gorenstein, Richard Lyons, and Ronald Solomon,** The classification of the finite simple groups, number 2, 1995

40.1 **Daniel Gorenstein, Richard Lyons, and Ronald Solomon,** The classification of the finite simple groups, number 1, 1994

39 **Sigurdur Helgason,** Geometric analysis on symmetric spaces, 1993

38 **Guy David and Stephen Semmes,** Analysis of and on uniformly rectifiable sets, 1993

37 **Leonard Lewin, Editor,** Structural properties of polylogarithms, 1991

36 **John B. Conway,** The theory of subnormal operators, 1991

35 **Shreeram S. Abhyankar,** Algebraic geometry for scientists and engineers, 1990

34 **Victor Isakov,** Inverse source problems, 1990

33 **Vladimir G. Berkovich,** Spectral theory and analytic geometry over non-Archimedean fields, 1990

32 **Howard Jacobowitz,** An introduction to CR structures, 1990

31 **Paul J. Sally, Jr. and David A. Vogan, Jr., Editors,** Representation theory and harmonic analysis on semisimple Lie groups, 1989

30 **Thomas W. Cusick and Mary E. Flahive,** The Markoff and Lagrange spectra, 1989

29 **Alan L. T. Paterson,** Amenability, 1988

28 **Richard Beals, Percy Deift, and Carlos Tomei,** Direct and inverse scattering on the line, 1988

27 **Nathan J. Fine,** Basic hypergeometric series and applications, 1988

26 **Hari Bercovici,** Operator theory and arithmetic in H^∞, 1988

25 **Jack K. Hale,** Asymptotic behavior of dissipative systems, 1988

24 **Lance W. Small, Editor,** Noetherian rings and their applications, 1987

23 **E. H. Rothe,** Introduction to various aspects of degree theory in Banach spaces, 1986

22 **Michael E. Taylor,** Noncommutative harmonic analysis, 1986

21 **Albert Baernstein, David Drasin, Peter Duren, and Albert Marden, Editors,** The Bieberbach conjecture: Proceedings of the symposium on the occasion of the proof, 1986

20 **Kenneth R. Goodearl,** Partially ordered abelian groups with interpolation, 1986

19 **Gregory V. Chudnovsky,** Contributions to the theory of transcendental numbers, 1984

18 **Frank B. Knight,** Essentials of Brownian motion and diffusion, 1981

17 **Le Baron O. Ferguson,** Approximation by polynomials with integral coefficients, 1980

16 **O. Timothy O'Meara,** Symplectic groups, 1978

15 **J. Diestel and J. J. Uhl, Jr.,** Vector measures, 1977

14 **V. Guillemin and S. Sternberg,** Geometric asymptotics, 1977

13 **C. Pearcy, Editor,** Topics in operator theory, 1974

12 **J. R. Isbell,** Uniform spaces, 1964

11 **J. Cronin,** Fixed points and topological degree in nonlinear analysis, 1964

10 **R. Ayoub,** An introduction to the analytic theory of numbers, 1963

(*Continued in the back of this publication*)

Symmetric Inverse Semigroups

MATHEMATICAL
Surveys and Monographs

Volume 46

Symmetric Inverse Semigroups

Stephen Lipscomb

American Mathematical Society
Providence, Rhode Island

Financial support from the following is gratefully acknowledged: Mary Washington College and the Navy-ASEE (American Society of Engineering Education) via the Marine Corps Systems Command Amphibious Warfare Directorate under the Marine Corps Exploratory Development Program (MQ1A PE 62131M).

1991 *Mathematics Subject Classification.* Primary 20M20, 20M18;
Secondary 20M05, 05C60, 20M30, 20B30.

ABSTRACT. With over 60 figures, tables, and diagrams, the text is both an intuitive introduction to and rigorous study of finite symmetric inverse semigroups. It turns out that these semigroups enjoy many of the classical features of the finite symmetric groups. For example, cycle notation, conjugacy, cummutativity, parity of permutations, alternating subgroups, Klein 4-group, Ruffini's result on cyclic groups, Moore's presentations of the symmetric and alternating groups, and the centralizer theory of symmetric groups are extended to more general counterparts in the semigroups studied here. We also classify normal subsemigroups, study congruences, and illustrate and apply an Eilenberg-style wreath product. The basic semigroup theory is further extended to partial transformation semigroups, and the Reconstruction Conjecture of graph theory is recast as a Rees ideal-extension conjecture.

Library of Congress Cataloging-in-Publication Data

Lipscomb, Stephen, 1944–
Symmetric inverse semigroups / Stephen Lipscomb.
p. cm. — (Mathematical surveys and monographs, ISSN 0076-5376 ; v. 46)
Includes bibliographical references and index.
ISBN 0-8218-0627-0 (alk. paper)
1. Inverse semigroups. I. Title. II. Series: Mathematical surveys and monographs ; no. 46.
QA182.L56 1996
512′.2—dc20
96-26968
CIP

∞ The paper used in this book is acid-free and falls within the guidelines
established to ensure permanence and durability.

10 9 8 7 6 5 4 3 2 1 01 00 99 98 97 96

dedicated to the memory of my father

David Leon Lipscomb (1913 – 1996)

Contents

Preface

"... *the invention of a convenient and flexible notation is often more seminal than is the proof of even the deepest of theorems.*" J. A. Paulos (1991)[1]

The primary purpose of this book is to document a study of the *symmetric inverse semigroup* C_n on the finite set $N = \{1, 2, \dots, n\}$. The semigroup C_n consists of all *charts* (one-one partial transformations) of N under the usual composition of mappings; and it contains the *symmetric group* S_n of permutations on N as a subgroup.

Over the last decade especially, certain aspects of the classical S_n theory have been naturally extended to C_n. For example, even permutations may now be viewed as special instances of even charts, an observation that led to the introduction of the alternating semigroups; studies of commutativity in C_n have unified the centralizer theories of C_n and S_n; and normal subsemigroups of C_n have been classified.

Part of the success of this S_n-to-C_n extension rests on the use of *path notation*, which may be illustrated by considering the chart

$$\alpha = \begin{pmatrix} 1 & 2 & 3 & 4 & 5 & 6 & 7 \\ 2 & 3 & - & 5 & 4 & 7 & - \end{pmatrix} \in C_7.$$

Expressed in path notation, $\alpha = (123](45)(67]$ has permutation part (45) and nilpotent part $(123](67]$.

A secondary purpose of this book is to document how path notation extends beyond C_n to the *partial transformation semigroup* PT_n on the finite set N. (The semigroup PT_n consists of all partial transformations of N under the usual composition of mappings.) The material on PT_n is of an introductory nature and appears in Chapters 11 and 12, where the results run parallel to the C_n theory appearing in Chapters 1 and 2.

The book is designed so that graduate students in either mathematics or computer science who have a basic knowledge of semigroups may proceed to original research in partial transformation semigroups.

Path notation, and subsequently certain sections of this book, grew out of my interest in the famous *Reconstruction Conjecture* of graph theory, where the *hypomorphic mapping sets* are sets of charts (*partial symmetries*). Quite briefly, I originally hoped that semigroup theory (especially the ideal

[1] [1], Page 163.

extension part) contained a solution to the Reconstruction Conjecture; but basically, I became frustrated because I could not multiply (compose) partial symmetries efficiently — there was no convenient and flexible notation for representing and multiplying charts, like cycle notation, which allows for representing and multiplying permutations (*full symmetries*).

It turned out, however, that cycle notation did extend to path notation, which works equally well for both partial and full symmetries. And, though the Reconstruction Conjecture remains a conjecture, path notation has proved useful in developing the C_n theory.

Much of the material for this text slowly evolved out of graduate and advanced-undergraduate courses that I taught for Virginia Technological & State University and Mary Washington College. In each course, I benefited greatly from students' criticisms and suggestions, especially those of Chris Dupilka, Margaret Hermann, and Dan Parks. (In fact, Parks [1] went on to introduce an application of inverse semigroup theory to quantum physics.)

Some of the research for this text was completed during my senior fellow appointment in the Navy-ASEE (American Society of Engineering Education) summer faculty research program; and for their support, I thank Mary Lacey and Robert Stiegler. I also thank those at Mary Washington College who provided me with various summer research grants (1985–1992).

Writing a mathematics book is very time consuming, and without mathematicians who share the necessary common background and willingness to provide feedback, it is perhaps impossible. So I give my deepest thanks to my colleague Janusz Konieczny, who carefully read the entire text and willingly spent many hours discussing its content. I also extend my gratitude to Mario Petrich, whose advisory comments led to a greatly improved presentation.

And a word of thanks to my wife Patty, who kept me healthy and happy.

May 1996 S.L.L.
Mary Washington College
Fredericksburg, Virginia

Introduction

In 1815, inspired by Gauss' "theory of forms" (in *Disquisitiones Arithmeticae*), Cauchy [1] published a memoir in which he introduced the (cycle) notation "(a, b)" to indicate the transposition of two letters a and b by a permutation. In the second part of that memoir, Cauchy [2] also introduced both the decomposition of a permutation into disjoint cycles and the alternating subgroup A_n of the symmetric group S_n.

Almost a century later (circa 1900), as an abstraction of permutation groups, various axiom systems for abstract group theory were being formulated and compared. Since that time, group theory has become more and more abstract, requiring less and less cycle notation. But throughout the history of group theory, and even today, cycle notation remains useful (if not fundamental) to the S_n theory.

In 1922, perhaps inspired by the evolution of group theory, A.K. Suschkewitsch essentially expressed the idea that the basic content of group theory is its relationship with transformation groups, and that the basic content of semigroup theory should be its relationship with transformation semigroups (Gluskin and Schein [1]). (A *transformation semigroup* on a set X is a semigroup whose elements are partial transformations on X and whose multiplication is the usual composition of functions.[1])

Thirty years later, in 1952, V.V. Wagner [1] was the first to introduce inverse semigroups, where the semigroup C_n of charts on $N = \{1, \ldots, n\}$ plays the role that S_n plays in the theory of groups. (Formally, a semigroup S is an *inverse semigroup* if it satisfies the additional property that for each $a \in S$, there is a unique $b \in S$ such that both $aba = a$ and $bab = b$.)

What we now call inverse semigroups, Wagner originally called *generalized groups*. These generalized groups were independently reintroduced in 1954 by G.B. Preston, who called them *inverse semi-groups*, and from the beginnings of the theory, Wagner and Preston recognized that the classical Cayley Theorem for groups had a semigroup analogue — each inverse semigroup is isomorphic to a subsemigroup of some symmetric inverse semigroup C_X of all charts on X. (This is the well-known *Wagner-Preston Theorem.*)

Over the next 30 years (1954–1984), the volume of literature concerning inverse semigroups grew substantially — by 1984, M. Petrich had published his 674-page text, *Inverse Semigroups*.

[1] A *partial transformation* on X is a function having both its domain and range included in X; and a *chart* on X is a one-one partial transformation on X.

During that same 30-year period, however, relatively little was written on symmetric inverse semigroups: For example, as early as 1953, A.E. Liber [1] had provided an algebraic characterization and studied congruences, but as late as 1994, the normal subsemigroups of C_n were yet to be classified.

The classification, published in 1995, is essentially a result of viewing C_n via a notation for representing its elements. To illustrate, consider that path notation yields the concept of an even chart, which, in turn, yields the normal subsemigroup A_n^c (alternating semigroup) of C_n.

As far as this author knows, from 1957 to 1987 there were four independently introduced notations for elements of C_n. The first appeared in a 1957 paper by W.D. Munn [1]; the second, path notation, in a 1986 paper by the author [1]; and the third and fourth in 1987 papers — one authored by G.M.S. Gomes and J.M. Howie [2], and another by R.P. Sullivan [1].

Format, Conventions, and Outline

The style of the text is informal, e.g., definitions of terms are generally neither numbered nor offset. A term that is being defined within a paragraph, however, always appears in italics. In contrast, lemmas, propositions, and theorems are always bold-face, numbered, and offset.

The sections are numbered sequentially throughout the text, from §**1** in Chapter 1, to §**79** (the last section) in the Appendix. Each figure, table, lemma, proposition, and theorem is numbered according to the section in which it appears. To illustrate, in §**51** we begin with "Figure 51.1." Then we have **51.2 Lemma**, which is followed by **51.3 Theorem**. In short, everything may be found by looking for the appropriate section numbers.

Throughout the text, S_n denotes the symmetric group of permutations on $N = \{1, 2, \ldots, n\}$, and C_n the symmetric inverse semigroup on N. Partial one-one transformations are called charts: A *chart* $\alpha \in C_n$ if and only if $\alpha : \mathbf{d}\alpha \to \mathbf{r}\alpha$ is a one-one function whose domain $\mathbf{d}\alpha$ and range $\mathbf{r}\alpha$ are subsets of N. Operators are usually written on the right, e.g., $i\alpha = j$, but sometimes they are also written on the left.

For specific contents of the chapters, let us consider them individually.

Chapter 1, *Decomposing Charts*: Cycle notation and the disjoint cycle decomposition of permutations are extended to path notation and the path decomposition of charts. The statement of the decomposition theorem is substantially motivated, and its proof is new.

Chapter 2, *Basic Observations*: In addition to yielding characterizations of idempotents, nilpotents, and inverse charts, path decomposition provides each chart $\alpha \in C_n$ with a permutation part γ and a nilpotent part η, i.e., $\alpha = \gamma\eta$. This "disjoint join" is often used to unify aspects of the S_n and C_n theories. For an example from this (most basic) chapter, the path structures of γ and η expose the cyclic subsemigroup theory of C_n, yielding Ruffini's "least common multiple of lengths of cycles" result in S_n as a corollary.

Similarly, conjugacy in S_n is shown to be corollary to conjugacy in C_n, i.e., charts are conjugate if and only if they have the "same path structure."

Chapter 3, *Commuting Charts*: The classical (cycle decomposition) characterization (circa 1870) of commuting permutations is extended (via path decomposition) to commuting charts.

Chapter 4, *Centralizers of Permutations*: Let α be a permutation in S_n. The classical isomorphism from $\{\beta \in S_n \mid \alpha \circ \beta = \beta \circ \alpha\}$ onto a "product of wreath products" is extended to a faithful representation of $C(\alpha) = \{\beta \in C_n \mid \beta \circ \alpha = \alpha \circ \beta\}$. The order of $C(\alpha)$ is calculated. These results derive from the path characterization of commutating charts (Chapter 3).

Chapter 5, *Centralizers of Charts*: The restriction that α be a permutation is removed — we consider the centralizer $C(\alpha)$ for an arbitrary chart $\alpha \in C_n$. The results unify the centralizer theories of S_n and C_n. The formula for the order of $C(\alpha)$ depends only on the path structure of α.

Chapter 6, *Alternating Semigroups*: The idea of an even permutation is generalized to charts; and the even charts in C_n form the alternating semigroup A_n^c. Generators of A_n^c are identified, and the T_{61} semigroup, the companion of the Klein 4-group, appears.

Chapter 7, *S_n-normal Semigroups*: The "permutation self-conjugate" subsemigroups of C_n are classified. The alternating semigroups play a significant role, and path notation is fundamental. The X-semigroups represent a "sporadic kind" of S_n-normal semigroup, i.e., semigroups which only appear when n is even. Lattice diagrams for the C_3 and C_4 cases are provided.

Chapter 8, *Normal Semigroups and Congruences*: Unlike the group theory case, where morphisms having domain S_n determine all normal subgroups of S_n, morphisms having domain C_n do not determine all normal subsemigroups of C_n. For example, the alternating semigroup $A_n^c \subset C_n$ is "non-morphism normal." Here, classical results on congruences of C_n combine with the material of Chapter 7 to provide complete knowledge of non-morphism normal subsemigroups of C_n.

Chapter 9, *Presentations of Symmetric Inverse Semigroups*: E. H. Moore's 1897 presentation of S_n is extended to one for C_n. The chapter opens with the necessary background on free algebras and presentations. The approach to finding a presentation for C_n is new and should have other applications. In fact, the same techniques are used in Chapter 10.

Chapter 10, *Presentations of Alternating Semigroups*: E. H. Moore's presentation of the alternating group A_n is extended to one for A_n^c.

Chapter 11, *Decomposing Partial Transformations*: If PT_n is the semigroup of partial transformations of $\{1, 2, \ldots, n\}$, then $S_n \subset C_n \subset PT_n$. It is therefore natural to extend path notation from C_n to PT_n. The extension is provided and used to study PT_n. The results obtained here run parallel to those of the C_n case (Chapters 1 and 2).

Chapter 12, *Commuting Partial Transformations*: Commuting Partial transformations $\alpha, \beta \in PT_n$ are characterized (in terms of path notation).

The characterization consists of three conditions, one characterizing commutativity in S_n, and two characterizing commutativity in C_n. An application yields a formula for the order of the centralizer of an idempotent in PT_n. Another application provides a short proof of one of Howie's theorems.

Chapter 13, *Centralizers, Conjugacy, Reconstruction*: In this final chapter, we review the state of the work on centralizers and pose a question first posed by Lallement. Conjugacy classes in C_n are counted, and Saito's (rather abstract) Theorem concerning the possibility of conjugacy (in C_n) of $\alpha \circ \beta$ and $\beta \circ \alpha$ is explained in the simple terms of path notation. The equivalence between "conjugacy" and "same path structure" that holds in both S_n and C_n is shown not to hold in PT_n. Certain aspects of graph theory are exposed as equivalent to aspects of semigroup theory. One result is that the foremost problem in graph theory, namely the Reconstruction Conjecture, is cast as an extension problem of one Brandt semigroup by another.

CHAPTER 1

Decomposing Charts

Following its fragmentary beginnings in the 1920s and 1930s, the algebraic theory of semigroups has grown from seminal attempts at generalizing group theory into a vast and independent branch of algebra. Of particular interest here is the extensively developed and exceptionally promising subbranch of *inverse semigroups*. Intuitively speaking, these semigroups are to "partial symmetry" what groups are to "symmetry."

It was immediately recognized (Wagner-Preston Theorem) that each inverse semigroup is isomorphic to a subsemigroup of a *symmetric inverse semigroup*. In the finite case, these symmetric inverse semigroups are denoted "C_n", $n = 1, 2, 3, \ldots$. Their members are called *charts* and the multiplication is function composition.[1] A chart $\alpha \in C_n$ if and only if $\alpha : \mathbf{d}\alpha \to \mathbf{r}\alpha$ is a one-one function whose domain $\mathbf{d}\alpha$ and range $\mathbf{r}\alpha$ are subsets of $N = \{1, 2, \ldots, n\}$. Since permutations of N are charts in C_n, the symmetric group S_n (of all permutations of N) is a subgroup of C_n.

Having S_n as a subgroup of C_n, we might suspect that the disjoint cycle decomposition of permutations somehow extends to charts, i.e., given any chart $\alpha \in C_n$, we desire to "decompose" $\alpha = \alpha_1 \cdots \alpha_k$ into certain "atomic charts" $\alpha_1, \ldots, \alpha_k$. In this chapter, we provide such a decomposition.

§1 Paths

In conjunction with the usual parentheses "(" and ")", path notation allows for the use of a right square bracket "]". The bracket "]" serves to specify those points that are not in the domain of a chart, e.g., $(1](2] \cdots (n]$ denotes the empty (or zero) chart $0 \in C_n$. Other examples are pictured below in Figure 1.1.

More precisely, for distinct elements $i_1, \ldots, i_k$ of N, let $\alpha \in C_n$ have domain $\mathbf{d}\alpha = \{i_1, \ldots, i_k\}$ and suppose $i_1\alpha = i_2$, $i_2\alpha = i_3$, $\ldots$, $i_{k-1}\alpha = i_k$, and $i_k\alpha = q$. Then α is a *path*. Turning on the value of q, we have two kinds of paths: If $q = i_1$ and $N - \mathbf{d}\alpha = \{j_1, \ldots, j_{n-k}\}$, then

$$\alpha = (i_1, i_2, \ldots, i_k)(j_1](j_2] \cdots (j_{n-k}]$$

[1] What we call charts are usually called *partial one-one transformations*. The choice of the term "chart" was motivated by "chart" as used in (topological) manifold theory.

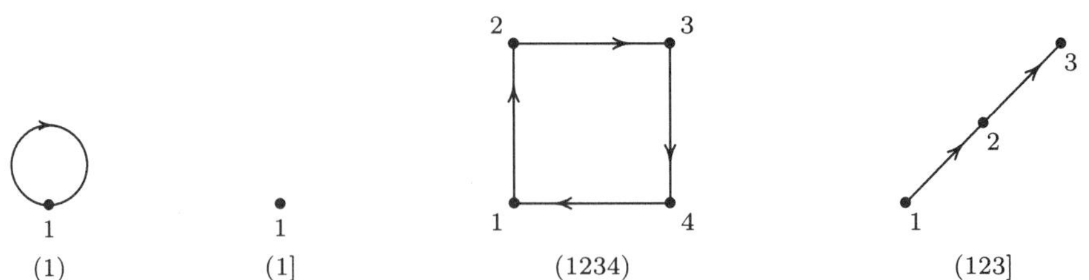

Figure 1.1. Picturing paths.

is a *circuit* (a k-circuit or a circuit of length k).[2] If $q \neq i_1$, then $N - \mathbf{d}\alpha = \{q, m_1, m_2, \ldots, m_{n-k-1}\}$ and

$$\alpha = (i_1, i_2, \ldots, i_k, q](m_1](m_2] \cdots (m_{n-k-1}]$$

is a *proper path* (a proper $(k+1)$-path or a proper path of length $k+1$).[3]

In addition to these paths (circuits of length ≥ 1 and proper paths of length ≥ 2), we define, for each $j \in N$, the *proper* 1*-path*

$$(j] = (1](2] \cdots (n] = 0 \in C_n.$$

We therefore have *ℓ-paths*, i.e., circuits and proper paths of *length* $\ell \geq 1$. For example, $(1] \cdots (i-1](i)(i+1] \cdots (n]$ denotes the 1-circuit with domain $\{i\}$, while $(12](3] \cdots (n]$ denotes the proper 2-path that maps 1 to 2. Every path has an obvious geometrical representation (Figure 1.1).

§2 Building Charts from Paths

To build charts from paths, let $\alpha, \beta \in C_n$ and suppose that $(\mathbf{d}\alpha \cup \mathbf{r}\alpha)$ and $(\mathbf{d}\beta \cup \mathbf{r}\beta)$ are disjoint. Then α and β are *disjoint* and the *join* γ of α and β (denoted $\gamma = \alpha\beta = \beta\alpha$) is the chart with domain $\mathbf{d}\alpha \cup \mathbf{d}\beta$ and values determined by

$$x\gamma = \begin{cases} x\alpha, & x \in \mathbf{d}\alpha \\ x\beta, & x \in \mathbf{d}\beta. \end{cases}$$

So the join $\gamma = \alpha\beta$ exists if and only if α and β are disjoint. For instance, the proper 2-path $\alpha = (12](3](4]$ and the 2-circuit $\beta = (1](2](34)$ are disjoint charts in C_4 and their join is $\gamma = \alpha\beta = (12](34)$. Note that we did not write $\gamma = (12](3](4](1](2](34)$, which would be confusing. It turns out that the explicit appearance of 1-paths "$(j]$" is often unnecessary. This is similar to

[2] Depending on context, the notation "$(i_1, \ldots, i_k)$" will denote either a circuit or a cycle.

[3] Depending on context, we use the notation "$(i_1, \ldots, i_k, q]$" to denote either a proper path or the chart $(i_1, \ldots, i_k, q](m_1) \cdots (m_{n-k-1})$.

the case of 1-cycles in cycle notation. To make matters worse, at times we shall also suppress 1-circuits "(j)."

Learning to multiply charts in path notation is like learning to multiply permutations in cycle notation, it takes a little practice. For starters, use the charts $\alpha = (123)(45]$ and $\beta = (41)(53)(2]$ in C_5 to calculate $\alpha \circ \beta = (1](25](34)$. Then practice taking powers of the proper 5-path $\gamma = (12345]$ — calculate that $\gamma^2 = (135](24]$, $\gamma^3 = (14](25](3]$, $\gamma^4 = (15](2](3](4]$, and $\gamma^5 = (1](2](3](4](5] = 0$.

§3 Decomposing Charts into Paths

Pick any chart $\alpha \in C_n$ and suppose that $x \in \mathbf{d}\alpha$. We shall form some proper paths and circuits that depend on the α-iterates of x: Let us look at the first iterate. We define

$$\eta_x = (x, x\alpha] \quad \text{if } x\alpha \neq x, \qquad \text{or} \qquad \gamma_x = (x) \quad \text{if } x\alpha = x.$$

Continuing with higher order iterates, for each $k \geq 2$, we also define

$$\eta_x = (x, x\alpha, x\alpha^2, \ldots, x\alpha^k] \quad \text{when } \{x, x\alpha, x\alpha^2, \ldots, x\alpha^k\} \text{ has size } k+1;$$
$$\gamma_x = (x, x\alpha, x\alpha^2, \ldots, x\alpha^{k-1}) \quad \text{when } \{x, x\alpha, x\alpha^2, \ldots, x\alpha^k = x\} \text{ has size } k.$$

Each η_x is a *proper path in* α; and each circuit γ_x is a *circuit in* α. But unlike circuits, each proper path $\eta = (i_1, i_2, \ldots, i_k]$ has a *left-endpoint* i_1 and a *right-endpoint* i_k. Moreover, a proper path η_x in α is *maximal* when its left endpoint $x \in \mathbf{d}\alpha - \mathbf{r}\alpha$ and its right endpoint $x\alpha^k \in \mathbf{r}\alpha - \mathbf{d}\alpha$.

To describe how the various paths in α must interact, we shall say that the path η *meets* the path γ whenever they are not disjoint, i.e., when

$$(\mathbf{d}\eta \cup \mathbf{r}\eta) \cap (\mathbf{d}\gamma \cup \mathbf{r}\gamma) \neq \emptyset.$$

To illustrate, note that the circuit (123) meets the proper 2-path $(43]$ at 3, while the proper paths $(1234]$ and $(5678]$ are disjoint.

3.1 Lemma

Let $\alpha \in C_n$. Then the following are true:

1. *For maximal paths η and η' in α, either $\eta = \eta'$ or η does not meet η'.*
2. *For circuits γ and γ' in α, either $\gamma = \gamma'$ or γ does not meet γ'.*
3. *No maximal path η in α meets any circuit γ in α.*
4. *For each $y \in \mathbf{r}\alpha - \mathbf{d}\alpha$, there exist $x \in \mathbf{d}\alpha - \mathbf{r}\alpha$ and $k \geq 1$ such that $x\alpha^k = y$, i.e., maximal $\eta_x = (x, x\alpha, \ldots, x\alpha^k = y]$ exists whenever $y \in \mathbf{r}\alpha - \mathbf{d}\alpha$.*

Proof. Statements (1)–(3) follow since $\alpha : \mathbf{d}\alpha \to \mathbf{r}\alpha$ is a chart. For (4), we use induction: Since $y \in \mathbf{r}\alpha - \mathbf{d}\alpha$, there exists $x_1 \in \mathbf{d}\alpha$ such that $x_1\alpha = y$ and $\{x_1, y\}$ has size two. If $x_1 \notin \mathbf{r}\alpha$, then, letting $x = x_1$ and $k = 1$, we see that $\eta_x = (x_1, y] = (x, x\alpha^1 = y]$ is maximal. Otherwise, there exists $x_2 \in \mathbf{d}\alpha$ such that $x_2\alpha = x_1$ and $\{x_2, x_1, y\}$ has size three. If $x_2 \notin \mathbf{r}\alpha$, then, letting $x = x_2$ and $k = 2$, we see that $\eta_x = (x_2, x_1, y] = (x, x\alpha^1, x\alpha^2 = y]$ is maximal. But construction (by induction) of such sets $\{x_k, x_{k-1}, \dots, x_1, y\}$ must terminate since N is finite. It follows that the desired x and k exist. □

Next, we have the fundamental decomposition/representation theorem.

3.2 Theorem (Unique Representation of Charts)

Every chart $\alpha \in C_n - \{0\}$ is a (disjoint) join

$$\eta_1 \cdots \eta_u \gamma_1 \cdots \gamma_v$$

of some (possibly none) length ≥ 2 proper paths $\eta_1, \dots, \eta_u$ and some (possibly none) circuits $\gamma_1, \dots, \gamma_v$. Moreover, this factorization is unique except for the order in which the paths are written.

Proof. If $\alpha \in C_n - \{0\}$, then either $\mathbf{d}\alpha = \mathbf{r}\alpha$ or $\mathbf{d}\alpha \neq \mathbf{r}\alpha$. In the former case, α is a permutation of its domain $\mathbf{d}\alpha$, and therefore has a disjoint "cycle" decomposition

$$\alpha = \delta_1 \circ \cdots \circ \delta_v \quad \text{(each } \delta_i \text{ is a cycle on } \mathbf{d}\alpha\text{)}.$$

For each index i, let γ_i be the restriction of δ_i to the set of points moved by δ_i. Then $\alpha = \gamma_1 \cdots \gamma_v$ is the desired decomposition. In the $\mathbf{d}\alpha \neq \mathbf{r}\alpha$ case, we may suppose that $\mathbf{d}\alpha - \mathbf{r}\alpha = \{1, 2, \dots, u\}$. Picking $x \in \mathbf{d}\alpha - \mathbf{r}\alpha$ and iterating, as far as possible, we obtain the set

$$X = \{x, x\alpha, x\alpha^2, \dots\}.$$

Since $X \subset N$ is finite, there exists a minimum $k \geq 1$ such that either $x\alpha^k \notin \mathbf{d}\alpha$ or $x\alpha^k = x\alpha^m$ for minimum $m > k$. But since the case $x\alpha^k = x\alpha^m$ is impossible because α is one-one, for each $x \in \mathbf{d}\alpha - \mathbf{r}\alpha$ there is a k such that $x\alpha^k \in \mathbf{r}\alpha - \mathbf{d}\alpha$ and we can define the maximal proper path

$$\eta_x = (x, x\alpha, \dots, x\alpha^k].$$

From (1) of 3.1, it is clear that we can define the join

$$\beta = \eta_1\eta_2 \cdots \eta_u.$$

Thus, if $\alpha = \beta$, we are finished. Otherwise, we shall show that a join of circuits $\alpha' = \gamma_1 \cdots \gamma_v$ exists such that $\alpha = \beta\alpha'$: First, define

$$\alpha' = \alpha|_{\mathbf{d}\alpha - \mathbf{d}\beta}$$

as the restriction of α to $\mathbf{d}\alpha - \mathbf{d}\beta$. To see that α' is a permutation, we shall need several properties:

(i) $\mathbf{r}\alpha - \mathbf{d}\alpha = \mathbf{r}\beta - \mathbf{d}\beta$ (Using (4) of 3.1, we have $y \in \mathbf{r}\alpha - \mathbf{d}\alpha \Leftrightarrow$ maximal $\eta_x = (x, \dots, y]$ exists for some $x \in \mathbf{d}\alpha - \mathbf{r}\alpha \Leftrightarrow y \in \mathbf{r}\beta - \mathbf{d}\beta$.)
(ii) $\mathbf{d}\beta - \mathbf{r}\beta = \mathbf{d}\alpha - \mathbf{r}\alpha$ (Both equal $\{1, 2, \dots u\}$.)
(iii) $\mathbf{r}\alpha' \cup \mathbf{r}\beta = \mathbf{r}\alpha$.

We can now show that $\mathbf{d}\alpha' \subset \mathbf{r}\alpha'$, implying that α' is a permutation of its domain $\mathbf{d}\alpha'$. Let $y \in \mathbf{d}\alpha'$. Then $y \in \mathbf{d}\alpha$ implies $y \notin \mathbf{r}\alpha - \mathbf{d}\alpha$. Coupling this with $y \notin \mathbf{d}\beta$ and (i), we deduce

(iv) $y \notin \mathbf{r}\beta$.

Similarly, from $y \notin \mathbf{d}\beta - \mathbf{r}\beta$ and $y \in \mathbf{d}\alpha$, (ii) shows

(v) $y \in \mathbf{r}\alpha$.

Then from (iv) and (v), property (iii) yields $y \in \mathbf{r}\alpha'$, which finishes the proof that α' is a permutation.

Next, for $x \in \mathbf{d}\alpha'$, we take α-iterates of x and get $X = \{x, x\alpha, x\alpha^2, \dots\}$. This time, there exists a $k \geq 1$ such that $x = x\alpha^k$: For each $x \in \mathbf{d}\alpha'$, define

$$\gamma_x = (x, x\alpha, \dots, x\alpha^{k-1}).$$

Then from (2) of 3.1, pairwise disjoint circuits $\gamma_1, \dots, \gamma_v$ exist such that $\alpha' = \gamma_1 \cdots \gamma_v$. Lastly, by (3) of 3.1, the paths in the factorization $\alpha = \beta\alpha' = \eta_1 \cdots \eta_u \gamma_1 \cdots \gamma_v$ are pairwise disjoint.

Turning to uniqueness, we suppose that

$$\alpha = \eta_1 \cdots \eta_u \gamma_1 \cdots \gamma_v = \eta_1' \cdots \eta_w' \gamma_1' \cdots \gamma_x'$$

has two such factorizations. If $u = 0$, then $\mathbf{d}\alpha = \mathbf{r}\alpha$, showing that $w = 0$. Thus, $\alpha = \gamma_1 \cdots \gamma_v = \gamma_1' \cdots \gamma_x'$ is a join of circuits, and the uniqueness follows from an induction argument (observe that each γ_i, as well as each γ_j', is a circuit in α and apply (2) of 3.1). So suppose $u \geq 1$. We claim that $u = w$: First, note that for any proper path η of length ≥ 2 we have $\eta = (1, 2, \dots, k] \Rightarrow \{1\} = \mathbf{d}\eta - \mathbf{r}\eta$. Therefore, since the decomposition $\eta_1 \cdots \eta_u$ of α is a join, $\mathbf{d}\alpha - \mathbf{r}\alpha$ has u elements and, likewise, since $\eta_1' \cdots \eta_w'$ is a join, $\mathbf{d}\alpha - \mathbf{r}\alpha$ has w elements, i.e., $u = w$. So by rearrangement, if necessary, we may assume that for each i $i =$ left endpoint of $\eta_i =$ left endpoint of η_i'. But then, since each η_i and each η_j' are maximal, an application of (1) of

3.1 shows $\eta_i = \eta_i'$ for each i. It therefore only remains to show that we can "cancel" the proper paths, i.e., show that

$$\gamma_1 \cdots \gamma_v = \gamma_1' \cdots \gamma_x',$$

which is case $u = 0$. To this end, assume that for some index i, $y \in \mathbf{d}\gamma_i$. Then $y \notin \mathbf{d}(\eta_1 \cdots \eta_u)$ because the decomposition is a join. This implies $y \in \mathbf{d}\gamma_j'$ for some index j and, consequently, $y\gamma_i = y\gamma_j'$, showing that the function $\gamma_1' \cdots \gamma_x'$ is an extension of $\gamma_1 \cdots \gamma_v$. By a similar argument, the latter is an extension of the former, and we are finished. □

From Theorem 3.2, each nonzero $\alpha \in C_n$ is a disjoint join

$$\alpha = (a_{11} \cdots a_{1k_1}] \cdots (a_{u1} \cdots a_{uk_u}](b_{11} \cdots b_{1m_1}) \cdots (b_{v1} \cdots b_{vm_v})$$

of proper paths of length ≥ 2 and circuits. If $\{j_1, \ldots, j_\ell\} = N - (\mathbf{d}\alpha \cup \mathbf{r}\alpha)$, then none of the j_i's appear in the representation specified in Theorem 3.2. We may, however, augment the Theorem 3.2 join with the proper 1-paths $(j_i]$ $(j_i \notin \mathbf{d}\alpha \cup \mathbf{r}\alpha)$ and obtain yet another unique representation. Indeed, augmenting the representation above, we obtain

$$\alpha = (j_1] \cdots (j_\ell](a_{11} \cdots a_{1k_1}] \cdots (a_{u1} \cdots a_{uk_u}](b_{11} \cdots b_{1m_1}) \cdots (b_{v1} \cdots b_{vm_v}),$$

which we shall call either the *path decomposition* or *join representation* of α. For instance, the decomposition of

$$\alpha = \begin{pmatrix} 1 & 2 & 3 & 4 & 5 & 6 & 7 \\ 2 & 3 & - & 5 & 4 & - & 7 \end{pmatrix} \in C_7$$

is (123](6](45)(7), which may be graphically represented as in Figure 3.3.

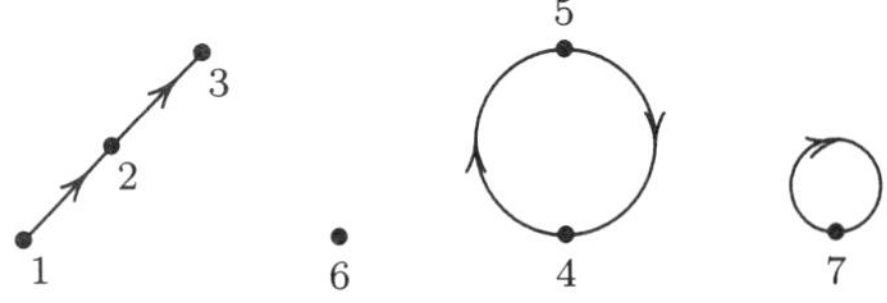

Figure 3.3. The path decomposition of a chart.

We also note that while the zero chart 0 of C_n is excluded from Theorem 3.2, it does have path decomposition $(1] \cdots (n]$. The zero $(1] \cdots (n]$ is an example of a *nilpotent*, which is a chart whose path decomposition contains no circuits. In fact, given $\alpha \in C_n$ with join representation above, its *nilpotent part* is

$$\eta = (j_1] \cdots (j_\ell](a_{11} \cdots a_{1k_1}] \cdots (a_{u1} \cdots a_{uk_u}](b_{11}] \cdots (b_{1m_1}] \cdots (b_{v1}] \cdots (b_{vm_v}];$$

and its *permutation part* is

$$\gamma = (j_1] \cdots (j_\ell](a_{11}] \cdots (a_{1k_1}] \cdots (a_{uk_u}](b_{11} \cdots b_{1m_1}) \cdots (b_{v1} \cdots b_{vm_v}).$$

In other words, each chart $\alpha = \eta\gamma$ is the join of its nilpotent and permutation parts. In particular, the chart $\alpha = (123](6](45)(7)$ pictured in Figure 3.3 has nilpotent part $\eta = (123](6] = (123](6](4](5](7]$ and permutation part $\gamma = (45)(7) = (45)(7)(1](2](3](6]$.

§4 Comments

As one might suspect, the literature on semigroups is rather diverse with certain of its areas extensively developed. Hille's book concerns the *analytic theory* of semigroups and its *applications to analysis*, while Birkhoff's text gives an account of *lattice-ordered* semigroups. On the other hand, the books by Suschkewitsch, Ljapin, and Clifford and Preston concern *algebraic semigroups* — those semigroups not endowed with any further structure.

Historically, it is claimed that the term "semigroup" first appeared in the mathematical literature in 1904 (page 8 of J.-A. de Séguier's book [1]), that the first published paper on semigroups appeared in 1905 (L. E. Dickson [1]), and that the first book on semigroups appeared in 1937 (A. K. Suschkewitsch [2]). (Clifford and Preston [1] and also Schein [1].)

From 1940 to 1961, according to Clifford and Preston, "... the number of papers [on semigroups] appearing each year has grown fairly steadily to a little more than 30 on average." Their estimate roughly equates to the 494 bibliographical entries in the 1958 (first) edition of Ljapin's book [1].

In 1952, Wagner introduced inverse semigroups as *generalized groups*, and two years later, in 1954, Preston independently discovered these semigroups, calling them *inverse semi-groups*. Subsequently, research activity in inverse semigroups was substantial: In 1984, M. Petrich published his 674-page text, *Inverse Semigroups*. It contains 546 bibliographical entries, 505 of which are dated after 1958, the year that Ljapin listed 494.

At the very beginnings of inverse semigroup theory, Wagner [1] in 1952, Preston [2] in 1954, and Preston [3] in 1957, proved the Wagner-Preston Theorem — each inverse semigroup is isomorphic to a subsemigroup of a symmetric inverse semigroup — the analogue of Cayley's Theorem.

Also in 1957, as part of his study of characters of symmetric inverse semigroups, W. D. Munn [1] was the first to discover a notational representation of charts that is essentially equivalent to path decomposition.

Munn's decomposition used "links" and "cycles," instead of proper paths and circuits. For example, given the chart $(18](29](345)(6)(7) \in C_9$, he would write $[18][29](345)(6)(7)$, the links being $[18]$ and $[29]$. Links were defined as sequences. Thus, for example, $[18]$ would be a map having domain of size 2.

In the context of path notation, however, (18] is a proper 2-path, having domain of size 1. Similarly, the 3-circuit "(345)" has domain of size 3, while in the context of Munn's notation, "(345)" is a cycle with domain of size 9. In spite of these differences, Munn's approach and the one used here yield essentially the same notational form.

By the mid-1980s, the idea of a proper path was evidently an idea waiting to happen: In 1986, the author [1] independently invented path notation (as presented here) and proved Theorem 3.2. (The approach grew out of a study of hypomorphic mapping sets in the famous Graph Reconstruction Conjecture (Chapter 13).) In the next year (1987), G. M. S. Gomes and J. H. Howie [2] independently introduced the notion of a *primitive nilpotent*, which they denoted "$||12\cdots k||$." (Unlike "links," primitive nilpotents are precisely proper paths.) In their Theorem 2.8, they show that a non-zero nilpotent in C_n is a disjoint union of primitive nilpotents, which is part of our Theorem 3.2. And also in 1987, R. P. Sullivan [1] independently defined *k-chains* "$[1,\dots,k+1]$" and *k-cycles* "$(1,\dots,k)$", which are, respectively, proper $(k+1)$-paths and k-circuits.

This mid-1980s idea of decomposing charts into paths merits comparison with the 1815 idea of decomposing permutations into cycles. In the permutation case, cycle decomposition appeared in 1815 along with the beginnings of finite group theory. In particular, in 1815 Cauchy [1, page 18] introduced cycle notation "(i,j)" for transpositions, factored a three cycle $(i,j,k)=(j,k)\circ(i,j)$, and then (Cauchy [2]) decomposed permutations into disjoint cycles. Cycle notation proved useful in the early (up to 1911) development of finite group theory: Burnside [1] opens his 1911 text *Theory of Groups of Finite Order* with the following comment on cycle notation,

> "AMONG the various notations used in the following pages, there is one of such frequent recurrence that a certain readiness in its use is very desirable in dealing with the subject of this treatise. We therefore propose to devote a preliminary chapter to explaining it in some detail."

Since 1911, however, the approach to group theory has become more and more abstract, requiring less and less cycle notation. Nevertheless, cycle notation remains useful, if not fundamental, to the S_n theory.

In contrast, conceived in the 1950s, inverse semigroup theory was axiomatic from its inception. It has the C_n theory as one of its branches and path notation did not appear until the mid-1980s. As to the state of the C_n theory in 1985, consider the following statement of Gomes and Howie [1]:

> "Since the theory of inverse semigroups is now extensive enough to have been the subject of a substantial book by Petrich [1], it is perhaps rather surprising that very little has been written on the symmetric inverse semigroup."

CHAPTER 2

Basic Observations

For the S_n theory, Cauchy's convenient and flexible cycle notation is a seminal, even fundamental, way of thinking. In a similar vein, for the C_n theory, path notation also provides a fundamental way of thinking.[1]

In this chapter, we present the path notation view of idempotent, nilpotent, and inverse charts. Conjugacy and path structure are introduced and connected, extending their group counterparts (from S_n to C_n). This connection is used to compute the conjugacy classes of C_4 (Table 7.1). Path notation is also applied to finite cyclic semigroups, thereby generalizing Ruffini's classical result [1] that the order of a permutation is the least common multiple of the order of its cycles. For completeness of the cyclic semigroup theory, we sharpen Frobenius' result [1].

§5 Idempotents, Nilpotents, Inverses, Conjugacy

First, recall that an element e in a semigroup S is an *idempotent* if $e^2 = e$. For example, each S_n, indeed any group, contains exactly one idempotent while each C_n contains several idempotents. For C_3 in particular, if i, j, k denote distinct members of $\{1, 2, 3\}$, then for idempotents we have one of rank 3, three of rank 2, three of rank 1, and one of rank 0:

rank 3	rank 2	rank 1	rank 0
$(1)(2)(3)$	$(i)(j)(k]$	$(i)(j](k]$	$(1](2](3]$

That each of these idempotents is a join of 1-paths is no accident.

Second, recall that an element a in a semigroup S with zero "0" is a *nilpotent* if one of its positive powers a^k is zero, i.e., $a^k = 0$ for some positive integer k. For example, each S_n $(n \geq 2)$ contains no nilpotents (since nontrivial groups contain no zero) while each C_n is rich in nilpotents. For C_3 in particular, we find nilpotents at several ranks, namely, six of rank 2, six of rank 1, and one of rank 0:

rank 3	rank 2	rank 1	rank 0
none	$(ijk]$	$(ij](k]$	$(1](2](3]$

[1]In the chapter *Path Notation for Partial Transformations*, we shall see how path notation extends from the symmetric inverse semigroups C_n to the more general partial transformation semigroups PT_n.

That these nilpotents are joins of proper paths is no accident.

Third, recall that a chart β is the *inverse* of a chart α if $\beta : \mathbf{r}\alpha \to \mathbf{d}\alpha$ is the inverse (function) of α. In such a case, we write $\beta = \alpha^{-1}$. As we shall see, "reversing paths" switches between a chart and its inverse, just as "reversing cycles" switches between a permutation and its inverse.

Finally, to extend (from S_n to C_n) the equivalence *Two permutations are conjugate if and only if they have the same cycle structure*, we need the following: For $\alpha, \beta \in C_n$, we say α is *conjugate* to β whenever there exists a permutation $\gamma \in S_n$ such that $\gamma^{-1}\alpha\gamma = \beta$. Moreover, if $\alpha = \alpha_1 \cdots \alpha_k$ and $\beta = \beta_1 \cdots \beta_m$ are join representations, then α and β have the *same path structure* whenever $k = m$ and there exists a permutation ϕ of $K = \{1, 2, \ldots, k\}$ satisfying, for each $i \in K$,

(i) length of α_i = length of $\beta_{i\phi}$, and
(ii) α_i is a proper path if and only if $\beta_{i\phi}$ is a proper path.

So $\alpha = (123](45](678)(9) = \alpha_1\alpha_2\alpha_3\alpha_4 \in C_9$ and $\beta = (67](489](1)(235) = \beta_1\beta_2\beta_3\beta_4 \in C_9$ have the same path structure because (the matching of paths) $\phi = (12)(34)$ satisfies (i) and (ii). Furthermore, we may use this ϕ to show that α is conjugate to β: First, as indicated below, place each α_i over $\beta_{i\phi}$ —

$$\begin{aligned} \alpha_1\alpha_2\alpha_3\alpha_4 &= (123](45](678)(9) \\ \beta_2\beta_1\beta_4\beta_3 &= (489](67](235)(1). \end{aligned}$$

Second, define $\gamma \in S_9$ via "first-to-second line vertical inspection," i.e, γ is given by $1 \mapsto 4$, $2 \mapsto 8$, $\ldots$. From the definition of γ, we have

$$\gamma^{-1}\alpha\gamma = \beta.$$

In other words, α moves i to j if and only if β moves $i\gamma$ to $j\gamma$. That α and β are conjugate whenever they have the same path structure is no accident.

5.1 Theorem

Let $\alpha, \beta, \eta \in C_n$ and let α have path decomposition $\eta_1 \cdots \eta_u\gamma_1 \cdots \gamma_v$ where each η_i is a proper path and each γ_i is a circuit. Then the following statements are true.

(1) *Chart α is an idempotent if and only if each path in the join representation of α has length one.*
(2) *Chart α is nilpotent if and only if each path in the join representation of α is a proper path.*
(3) *The decomposition of α^{-1} is $\eta_1^{-1} \cdots \eta_u^{-1}\gamma_1^{-1} \cdots \gamma_v^{-1}$ where for proper paths, $(i_1, \ldots, i_k]^{-1} = (i_k, \ldots, i_2, i_1]$, and for circuits, $(j_1, \ldots, j_k)^{-1} = (j_k, \ldots, j_2, j_1)$.*
(4) *Chart α is conjugate to chart β if and only if α and β have the same path structure.*

(5) *For a proper path* $\eta = (i_1, i_2, \dots, i_k]$ *of length* $k \geq 2$, *we have* $\eta^k = 0$ *but, for each positive integer* $m < k$, $\eta^m \neq 0$.

(6) *For any nonzero integer* k, *we may calculate the* k*th power of* α *using* $\alpha^k = \eta_1^k \cdots \eta_u^k \gamma_1^k \cdots \gamma_v^k$.

Proof. To prove (1), use $i\alpha \neq i$, for $i \in \mathbf{d}\alpha$, implies $i\alpha^2 \neq i\alpha$. To prove (2), observe that $\alpha = \eta\gamma$ has a permutation part γ if and only if $\alpha^n \neq 0$ for every $n \geq 1$. To prove (3), use $i\alpha = j$ if and only if $j\alpha^{-1} = i$, and then recall that the path decomposition of α^{-1} is unique. For (4), first show that conjugates α and β necessarily have the same path structure by observing that any conjugate of an ℓ-circuit is an ℓ-circuit, and any conjugate of a proper ℓ-path is a proper ℓ-path. Second, show that charts α and β having the same path structure are conjugate by formalizing the argument given in the paragraph preceding this theorem. To prove (5), observe that $1 \leq m < k$ implies $i_1\eta^m = i_{m+1}$, and, since $\eta^k = \eta^{k-1}\eta$, that the domain $\mathbf{d}(\eta^k)$ of η^k is empty. To prove (6), recall that the paths appearing in the decomposition of α are pairwise disjoint. □

In light of Theorem 5.1, it is instructive to verify that C_n is an inverse (and hence regular) semigroup. Recall that a semigroup S is an *inverse semigroup* if every element has a unique inverse, i.e., for each $a \in S$, there exists a unique solution $x \in S$ of the system $axa = a$, $xax = x$. An equivalent definition is that S is a regular semigroup whose idempotents commute. (A semigroup S is *regular* if each of its elements is regular, i.e., for each $a \in S$, there exists $x \in S$ such that $axa = a$.)

§6 Index and Period of a Chart

If a is any element in any semigroup S, then the *cyclic subsemigroup* $\langle a \rangle$ of S *generated by* a consists of all positive integral powers of a. The *order* of a is the order of $\langle a \rangle$. Since we shall only consider finite cyclic semigroups $\langle a \rangle$, there exist positive integers r and s with $r < s$ such that $a^r = a^s$. Moreover, if s is the smallest such positive integer, then $a, a^2, \dots, a^{s-1}$ are distinct and r is the only integer less than s with $a^r = a^s$. The integer r is called the *index* and $m = s - r$ the *period* of a (of $\langle a \rangle$). We always have *index* + *period* = *order* + 1, and the subsemigroup $K_a = \{a^r, a^{r+1}, \dots, a^{r+m-1}\}$ of S is a cyclic group of order m. Each finite $\langle a \rangle$ contains exactly one idempotent, namely, the identity element of K_a.

The key relation between $\langle a \rangle$ and C_n is illustrated in Figure 6.1, which pictures an isomorphism between an abstract cyclic semigroup $\langle a \rangle$, of index 3 and period 4, and $\langle (123](4567) \rangle \in C_7$.

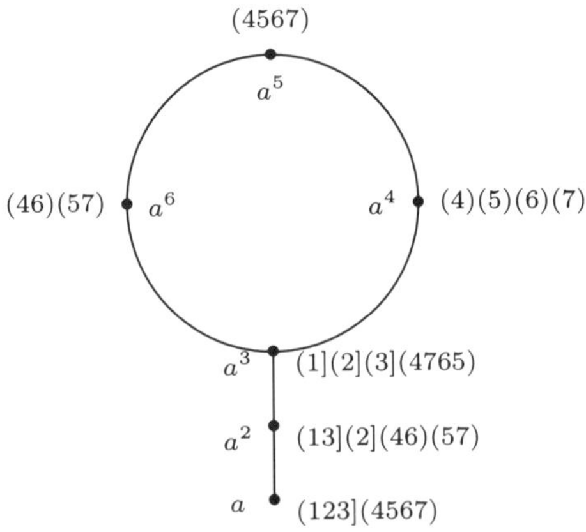

Figure 6.1. A cyclic semigroup isomorphism.

In addition to such isomorphisms, the ideas developed below are best motivated by simple examples: Calculate the period of $\alpha = (12)(345)(6789] \in C_9$ and the index of $\eta = (12](345](6789]$, the former being the least common multiple of 2 and 3, while the latter is the maximum of 2, 3, and 4. Then calculate the idempotents $\alpha^6 = (1)(2)(3)(4)(5)(6](7](8](9]$ and $\eta^4 = 0 \in C_9$.

6.2 Proposition

Let $\langle a\rangle$ be any finite cyclic semigroup with index r and period m. Then

(1) *the cyclic semigroup $\langle a\rangle$ is isomorphic to the cyclic subsemigroup $\langle\,(1,2,\ldots,r](r+1,r+2,\ldots,r+m)\,\rangle \subset C_{r+m}$ generated by $\alpha = (1,2,\ldots,r](r+1,\ldots,r+m)$; and*
(2) *the smallest exponent k such that a^k is the identity of K_a is the smallest $k \geq r$ that is also a multiple of m.*

Proof. Suppose $\langle a\rangle$ has index r and period m. To see (1), first calculate that $\langle(1,2,\ldots,r](r+1,\ldots,r+m)\rangle$ has index r and period m. Then observe that $a \mapsto (1,2,\ldots,r](r+1,\ldots,r+m)$ induces the desired isomorphism. To see (2), apply the isomorphism: For each $k = 1,2,\ldots,$ we see that a^k corresponds to $\alpha^k = (1,2,\ldots,r]^k(r+1,\ldots,r+m)^k$. By (1) of Theorem 5.1, α^k is an idempotent if and only if α^k is a join of 1-paths. So k is an exponent such that a^k is the identity of K_a if and only if

$$(1,2,\ldots,r]^k = 0 = (1](2]\cdots(r] \quad \text{and}$$
$$(r+1,r+2,\ldots,r+m)^k = (r+1)(r+2)\cdots(r+m).$$

But the first equation is equivalent to $k \geq r$, while the second is equivalent to k being a multiple of m. Thus, the smallest exponent k such that a^k is the identity of K_a is also the smallest $k \geq r$ that is also a multiple of m. □

6.3 Theorem

Let $\alpha \in C_n$. Then the following statements are true.

(1) *The index of α is the maximum of the lengths of the proper paths appearing in its join representation. If no proper paths appear, then its index is one.*
(2) *The period of α is the least common multiple of the lengths of the circuits appearing in its join representation. If no circuits appear, then its period is one.*

Proof. For (1), if no proper paths appear in the join representation of $\alpha \in C_n$, then $\alpha \in S_n$ is a permutation and the index of α is obviously one while the order of α is just the period of α. That the index of an $\alpha \in C_n - S_n$ can be calculated as claimed, we merely apply (5) and (6) of Theorem 5.1. For the proof of (2), suppose α has an ℓ-circuit γ in its decomposition. Then, since γ^k can be an idempotent only when k is a multiple of ℓ, it follows from (6) of 5.1 that the period of α is the least common multiple of the lengths of the circuits appearing in the decomposition of α. On the other hand, if no circuits appear, then α^k is 0 for some k and the period is one. □

If $\alpha \in S_n$, then (2) of 6.3 specializes to Ruffini's result: *The order of a permutation is the least common multiple of the lengths of its cycles.*

6.4 Corollary

Let $\langle a \rangle$ be a cyclic semigroup with index r and period m. If $q = \lfloor (r-1)/m \rfloor$ is the integer part of $(r-1)/m$, then $k = (q+1)m$ is the smallest integer such that a^k is the identity in K_a.

Proof. From 6.2, it suffices to show that k is the smallest multiple of m such that $k \geq r$. First, $k - m = qm < r$: This follows from

$$qm = \left\lfloor \tfrac{r-1}{m} \right\rfloor m \leq \tfrac{r-1}{m} m = r - 1 < r.$$

Second, $k = qm + m \geq r$: Now $r - 1 = \ell m + t$ for some t, $0 \leq t < m$. Then,

$$\left\lfloor \tfrac{r-1}{m} \right\rfloor = \left\lfloor \tfrac{\ell m + t}{m} \right\rfloor = \left\lfloor \ell + \tfrac{t}{m} \right\rfloor = \ell = \tfrac{r-1-t}{m}, \quad \text{showing that}$$

$$k = qm + m = \left\lfloor \tfrac{r-1}{m} \right\rfloor m + m = \tfrac{r-1-t}{m} m + m = r - 1 - t + m \geq r,$$

which finishes the proof. □

In the next section, we consider the cyclic subsemigroups of C_4.

§7 Cyclic Subsemigroups of C_n

Expressing $\alpha \in C_n$ in its join representation, we "know" (via Theorem 6.3) the index and period of α, i.e., we know $\langle \alpha \rangle$ up to isomorphism. As a consequence, the isomorphism classes of cyclic subsemigroups of C_n can be determined by listing all possible forms of join representations.

In addition, a conjugacy class of C_4 is just the set of all charts in C_4 having a given path structure. It follows that each "path structure form"

$$(\cdots]\cdots(\cdots](\cdots)\cdots(\cdots)$$

determines, and is determined by, a conjugacy class.

Table 7.1. Path structure of charts in C_4.

number of $m \to m$ maps	**path structure**	**number**	**index**	**period**
$\binom{4}{0}^2 0! = 1$	(1](2](3](4]	1	1	1
$\binom{4}{1}^2 1! = 16$	(1)(2](3](4]	4	1	1
	(12](3](4]	12	2	1
$\binom{4}{2}^2 2! = 72$	(1)(2)(3](4]	6	1	1
	(12)(3](4]	6	1	2
	(12](3)(4]	24	2	1
	(123](4]	24	3	1
	(13](24]	12	2	1
$\binom{4}{3}^2 3! = 96$	(1)(2)(3)(4]	4	1	1
	(12)(3)(4]	12	1	2
	(123)(4]	8	1	3
	(12](3)(4)	12	2	1
	(12](34)	12	2	2
	(123](4)	24	3	1
	(1234]	24	4	1
$\binom{4}{4}^2 4! = 24$	(1)(2)(3)(4)	1	1	1
	(12)(3)(4)	6	1	2
	(123)(4)	8	1	3
	(1234)	6	1	4
	(12)(34)	3	1	2
	Total =	**209**		

Thus, since Table 7.1 contains all possible path structure forms for charts in C_4, we conclude that C_4 has 20 conjugacy classes. As to the size of each of these conjugacy classes, we need only look under the column headed "number." And using the columns headed "index" and "period," we see that these conjugacy classes combine to form eight isomorphism classes. Recall that the number of charts in C_4 is

$$\sum_{m=0}^{4} \binom{4}{m}^2 m! = 209.$$

§8 Comments

Both finite group theory and cycle notation date from the time of Cauchy. For the development of finite group theory prior to 1911, see W. Burnside's text [1], and for many of the developments after 1911, see any standard reference, e.g., Suzuki [1] and [2]. As for cycle notation, inspired by Gauss' use of "the theory of forms," Cauchy introduced the notation in his 1815 memoir (Cauchy [1] and [2]).

One benefit of cycle notation is that in S_n, cycle structure determines conjugacy. This result is rather old, appearing as a "first principle" in §32 of C. Jordan's 1870 text [1], while its (obvious) extension to C_n, that path structure determines conjugacy, is rather new, introduced in 1992 by the author [5]. By 1995, the more general view of conjugacy was fundamental to classifying the S_n-normal semigroups (Chapter 7).

As for conjugacy in arbitrary semigroups, there seems to be no standard. One definition is given in Dauns [1], where semigroups with a zero have every pair of elements conjugate, two equivalent definitions appear in Lallement [1], which concern words in a free semigroup, and yet another in Goldstein and Teymouri [1], which is a modification of the one used by Lallement. Moreover, when the two equivalent definitions of conjugacy used by Lallement are extended to arbitrary monoids, they become different concepts (Zhang [1] and [2]).

While 3.1 extends to C_n the unique representations enjoyed in S_n, Theorem 5.1 extends to C_n the well known "algebra of cycle notation" that allows one to "see (or anticipate) the basics." For example, consider the following:

(i) With (1) of 5.1, which says that idempotents are joins of 1-paths, we shall "see" in §10 that idempotents in any finite inverse semigroup commute.

(ii) With (2) of 5.1, which says that nilpotents are joins of proper paths, we "see" the following proposition because we "essentially see" the proper paths "moving A to the right," and not "fixing A," which would require circuits in the representation of α.

> A chart $\alpha \in C_n$ of rank $k < n$ is nilpotent if and only if there exists no nonempty subset A of $\mathbf{d}\alpha$ such that $A\alpha = A$.

(iii) With (3) and (4) of 5.1, we "see" that α^{-1} and α are conjugate charts. Moreover, we may easily find $\beta \in S_n$ such that $\beta^{-1}\alpha\beta = \alpha^{-1}$, which would be rather tedious without the notation.

(iv) With (5) and (6) of 5.1, we have pictures such as Figure 6.1, where we "see," or may easily predict, the behavior of powers of charts without performing any multiplications.

For idempotents in other semigroups, in 1966 Howie [2] published a seminal paper where he characterized the elements of T_X, the semigroup (under composition) of all total transformations $X \to X$, that may be written as a product of idempotents in T_X. His work was extended to PT_X, the semigroup (under composition) of all partial (not necessarily one-one) transformations $X \to X$, for finite X in 1969 by Sullivan [2], and, independently, in 1970 and 1972 for arbitrary X by Evseev and Podran [1] and [2]. For other examples and information on research in this area, see Howie's 1980 and 1984 papers [3] and [4], see the introduction of Sullivan's 1987 paper [1], and Saito's 1989 papers [1] and [2].

Nilpotents do not appear as elements of groups because nontrivial groups have no zero. If a finite transformation semigroup S contains a zero, then it contains nilpotents, and it is natural to ask about the subsemigroup of S generated by all the nilpotents in S. In 1987, Gomes and Howie [2] and Sullivan [1] independently initiated this study for C_n and PT_n, respectively. (Recall that PT_n denotes the semigroup of all partial (not necessarily one-one) transformations $N \to N$.) Nilpotents have also been studied in other special cases, e.g., for a recent (1994) reference see Garba [1].

The notion of regularity was introduced into ring theory in 1936 by J. von Neumann. In the general theory of semigroups, regular semigroups were first studied by Thierrin [1] in 1951 under the name of *demi-groupes inversifs*.

Regular semigroups with commuting idempotents became important in 1952 when Wagner [1] introduced inverse semigroups. In 1952 and 1954, respectively, Wagner [1] and Preston [1] independently showed that a regular semigroup with commuting idempotents is an inverse semigroup. In 1953, Liber [1] proved the converse — any inverse semigroup is necessarily regular with commuting idempotents. Equivalent results were obtained in 1955 by Munn and Penrose [1].

It was G. Frobenius [1] in 1895 who first characterized the smallest exponent k such that a^k is the identity of K_a as the positive integer k satisfying $m|k$ and $r \le k < r + m$. That every element a in every finite semigroup determines a k such that a^k is idempotent was also proved in 1902 by E. H. Moore [2]. Corollary 6.4, which provides a formula $k = k(r, m)$ for calculating k, was discovered by the author [1]. The results of Frobenius were also found by Morgan Ward in 1933 (unpublished); Suschkewitsch [2, §19 of Chapter 2] (1937); Poole [1] (1937); Rees [1] (1940); and Climescu [1] (1946).

In the S_n theory, it is standard practice (Rotman [1, pages 41–44]) to use cycle notation to catalog conjugacy classes and/or count cyclic subgroups. The extension of such catalogs to include C_n (Table 7.1) was natural — path structure determines both the conjugacy classes in C_n and the isomorphism classes of cyclic subsemigroups of C_n (Theorem 6.3).

CHAPTER 3

Commuting Charts

We generalize the classical (circa 1870) characterization of commuting permutations — if $\alpha, \beta \in S_n$ and α has disjoint cycle decomposition $\gamma_1 \cdots \gamma_v$, then $\alpha \circ \beta = \beta \circ \alpha$ if and only if β maps each γ_i onto a γ_j of the same length, preserving the α-induced cyclic ordering. The generalization necessarily concerns both the nilpotent and permutation parts of a chart — if $\alpha, \beta \in C_n$ and α has path decomposition $\eta_1 \cdots \eta_u \gamma_1 \cdots \gamma_v$, then $\alpha \circ \beta = \beta \circ \alpha$ if and only if β maps some (possibly none) of the circuits γ_i onto some of the γ_i, preserving both length and α-induced cyclic ordering, and, in addition, β also maps some (possibly none) *initial segments* of the η_j onto some *terminal segments* of the η_j, preserving the α-induced linear ordering.

We begin with examples of mappings of circuits and segments (§9). Following the examples, we formalize the idea of initial and terminal segments of a proper path, and then make precise the concept of mapping the former onto the latter (§10). These notions are fundamental to understanding our characterization of commuting charts (Theorem 10.1), which will be used to unify the centralizer theories of S_n and C_n (Chapters 4 and 5).

§9 Examples

If $\alpha = (123)(45)(678)$ and $\beta = (173628)(4)(5)$ are permutations in S_8, then $\alpha \circ \beta = (182637)(45) = \beta \circ \alpha$. We note that β not only permutes the disjoint 3-cycles, (123) and (678), but also, in doing so, preserves their α-induced cyclic orderings (Figure 9.1).

$$\begin{array}{ll} \alpha = & (\ 1\ 2\ 3\ )\ (\ 4\ 5\ )\ (\ 6\ 7\ 8\ ) \\ & \quad\ \downarrow\ \downarrow\ \downarrow \quad\ \ \downarrow\ \downarrow \quad\ \ \downarrow\ \downarrow\ \downarrow \quad \downarrow \beta \\ \alpha = & (\ 7\ 8\ 6\ )\ (\ 4\ 5\ )\ (\ 2\ 3\ 1\ ) \end{array}$$

Figure 9.1. Picturing commuting permutations.

The C_n case is similar. For the same permutation $\alpha = (123)(45)(678)$ and the restriction $\beta' = (17](28](36] \in C_8$ of β, we again have $\alpha \circ \beta' = (18](26](37] = \beta' \circ \alpha$. In this case, however, β' maps only the 3-circuit (123) onto the 3-circuit (678) (Figure 9.2).

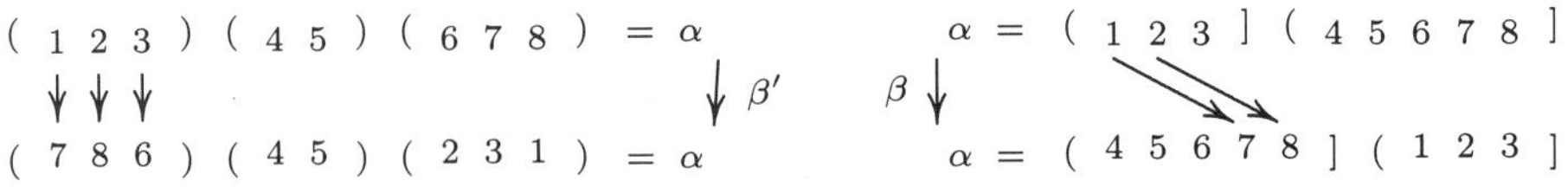

(1 2 3) (4 5) (6 7 8) $= \alpha$ $\qquad$ $\alpha =$ (1 2 3] (4 5 6 7 8]

↓ ↓ ↓ $\qquad$ ↓ β' $\qquad$ β ↓

(7 8 6) (4 5) (2 3 1) $= \alpha$ $\qquad$ $\alpha =$ (4 5 6 7 8] (1 2 3]

Figure 9.2. Picturing commuting charts.

For our final example, we stay in C_8 but change both α and β — we let $\alpha = (123](45678]$ and $\beta = (17](28]$. Then $\alpha \circ \beta = (18] = \beta \circ \alpha$, and we note that β, while preserving the α-induced ordering, maps the initial segment "(12" of (123] onto the final segment "78]" of (45678] (Figure 9.2).

§10 Characterization of Commuting Charts

Let $k \geq 2$ and suppose that $\eta = (i_1 \cdots i_k]$ is a proper k-path. Then η has

$$\begin{aligned} \textit{initial sets}: &\ \{i_1\}, \{i_1, i_2\}, \ldots, \{i_1, i_2, \ldots, i_k\}; \\ \textit{terminal sets}: &\ \{i_1, i_2, \ldots, i_k\}, \ldots, \{i_{k-1}, i_k\}, \{i_k\}; \\ \textit{initial segments}: &\ (i_1], (i_1 i_2], \ldots, (i_1 \cdots i_k]; \quad \text{and} \\ \textit{terminal segments}: &\ (i_1 \cdots i_k], \ldots, (i_{k-1} i_k], (i_k]. \end{aligned}$$

For $k = 1$, any attempt to define these concepts becomes a sticky wicket, because $(1] = (2] = \cdots = (n] = 0 \in C_n$. Nevertheless, we shall say that the *initial* and *terminal set* of "$(i]$" is the singleton set $\{i\}$, keeping in mind that (in this case) $\{i\}$ in determined by the representation "$(i]$" of the chart $0 \in C_n$. Similarly, for each (representation) "$(i]$", we shall say that its *initial* and *terminal segment* is $(i]$.

In order to define the concept of mapping segments onto segments, let β and the proper path $\eta = (i_1 \cdots i_k]$ be members of C_n. If there is an index w such that $\mathbf{d}\beta$ meets $\{i_1, \ldots, i_w, \ldots, i_k\}$ in an initial set $\{i_1, \ldots, i_w\}$, then for any proper path $\eta' = (j_1 \cdots j_v \ell_1 \cdots \ell_w]$ $(v \geq 0)$ such that $i_1\beta = \ell_1, \ldots, i_w\beta = \ell_w$, we shall say that *$\beta$ maps an initial segment of η onto a terminal segment of η'*. This concept is used in the following theorem.

10.1 Theorem (Commuting Charts)

Let $\alpha, \beta \in C_n$, and let α have the path decomposition

$$\alpha = (a_{11} \cdots a_{1r_1})(a_{21} \cdots a_{2r_2}) \cdots (a_{v1} \cdots a_{vr_v})(c_{11} \cdots c_{1p_1}] \cdots (c_{u1} \cdots c_{up_u}].$$

Then $\alpha \circ \beta = \beta \circ \alpha$ if and only if the path decomposition of α may be written

$$\alpha = (b_{11} \cdots b_{1r_1})(b_{21} \cdots b_{2r_2}) \cdots (b_{v1} \cdots b_{vr_v})(d_{11} \cdots d_{1q_1}] \cdots (d_{u1} \cdots d_{uq_u}]$$

where we have the following: (1) if $a_{ij} \in \mathbf{d}\beta$, then $a_{i1}, \ldots, a_{ir_i} \in \mathbf{d}\beta$ and $a_{i1}\beta = b_{i1}, \ldots, a_{ir_i}\beta = b_{ir_i}$, and (2) if $c_{ij} \in \mathbf{d}\beta$, then, for some p, $1 \leq$

$p \le p_i$, and some q, $1 \le q \le q_i$, $\{c_{i1}, \dots, c_{ip}\} = \mathbf{d}\beta \cap \{c_{i1}, \dots, c_{ip_i}\}$ and $c_{i1}\beta = d_{iq}$, $c_{i2}\beta = d_{i,q+1}$, $\dots$, $c_{ip}\beta = d_{iq_i}$.

Proof. Let $x, y \in N = \{1, 2, \dots, n\}$ and suppose $\delta \in C_n$. We shall use $x \xrightarrow{\delta} y$ to mean that $x \in \mathbf{d}\delta$ and that $x\delta = y$. Our proof is based on the observation that $\alpha \circ \beta = \beta \circ \alpha$ is equivalent to

and

$$\begin{array}{ccc} x & & \\ \beta\downarrow & & \\ w & \xrightarrow[\alpha]{} & z \end{array} \qquad \text{extends to} \qquad \begin{array}{ccc} x & \xrightarrow{\alpha} & y \\ \beta\downarrow & & \downarrow\beta \\ w & \xrightarrow[\alpha]{} & z. \end{array}$$

Figure 10.2. Expressing $\alpha \circ \beta = \beta \circ \alpha$ with commuting diagrams.

First, suppose $x\beta$ is in a circuit of α, i.e., assume $x\beta = b_{11}$. Then $(x\beta)\alpha$, $(x\beta)\alpha^2$, $\dots$, $(x\beta)\alpha^{r_1}$ are all defined. This data is organized in Figure 10.3.

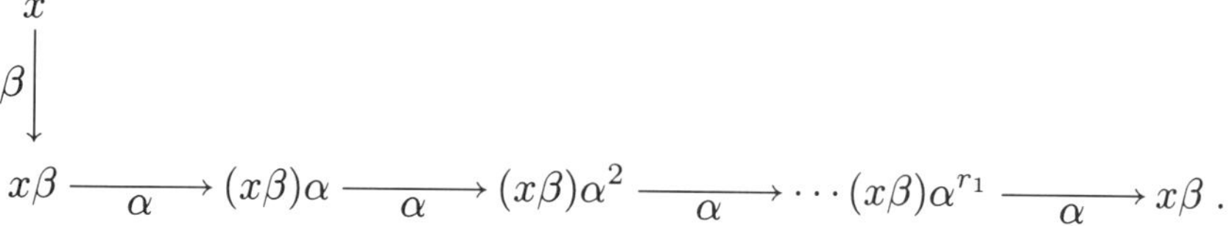

Figure 10.3. When $x\beta$ is in a circuit of α.

But viewing $\alpha \circ \beta = \beta \circ \alpha$ in terms of the diagrams in Figure 10.2, we may complete the diagram in Figure 10.3 as indicated in Figure 10.4.

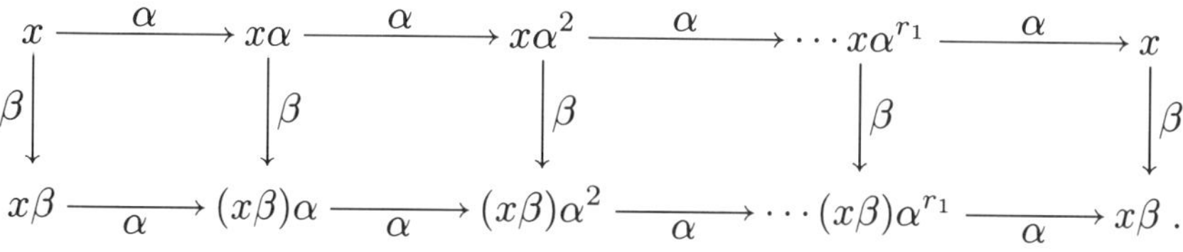

Figure 10.4. Completing the diagram in Figure 10.3.

It follows, since β is one-one, that x is also in an r_1-circuit of α. So without loss of generality, we may assume that $x = a_{11}$. Then, since β is one-one, an induction argument on the number v of circuits shows that (1) is true.

Second, assume that $x\beta$ is in a proper path of α, say $x\beta = d_{1j} \in \{d_{11}, \dots, d_{1q_1}\}$. In this second case, there exists a k, $0 \le k \le q_1$, such that the sequence $x\beta = (x\beta)\alpha^0 = d_{1j}$, $(x\beta)\alpha = d_{1,j+1}, \ldots, (x\beta)\alpha^k = d_{1q_1}$ terminates at $(x\beta)\alpha^k = d_{1q_1} \notin \mathbf{d}\alpha$ (Figure 10.5).

$$\begin{array}{l} x \\ \beta\big\downarrow \\ x\beta \xrightarrow[\alpha]{} (x\beta)\alpha \xrightarrow[\alpha]{} (x\beta)\alpha^2 \xrightarrow[\alpha]{} (x\beta)\alpha^3 \cdots \xrightarrow[\alpha]{} (x\beta)\alpha^k \ . \end{array}$$

Figure 10.5. When $x\beta$ is in a proper path of α.

As before, viewing $\alpha \circ \beta = \beta \circ \alpha$ in terms of the diagrams in Figure 10.2, we may fill out the diagram in Figure 10.5 as indicated in Figure 10.6.

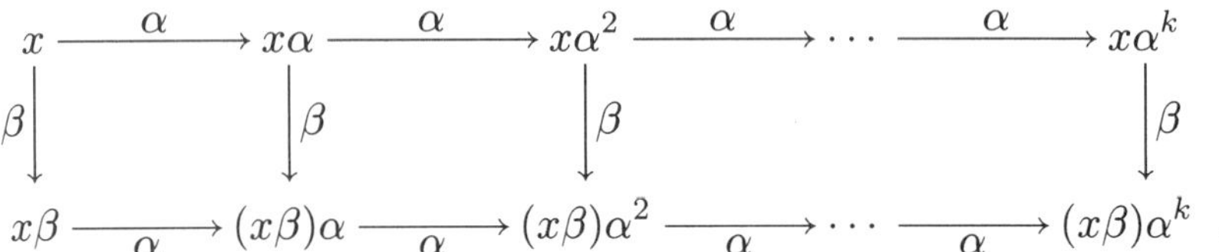

Figure 10.6. Completing the diagram in Figure 10.5.

Next, observe that $x\alpha^k$ must be in a proper path. (Otherwise, $x\alpha^k$ is in a circuit and then an $m \geq 1$ exists such that $(x\alpha^k)\alpha^m = x\alpha^{k+m} = x$. This implies that $x\beta = (x\alpha^{k+m})\beta = (x\beta\alpha^k)\alpha^m$, which contradicts $x\beta\alpha^k \notin \mathbf{d}\alpha$.) So we may suppose that $x \in \{c_{11}, \ldots, c_{1p_1}\}$. To see that β satisfies (2) relative to this proper path, we consider two cases: (*Case* I) If $x = c_{11}$, then, letting $p = k+1$ and defining q via $x\beta = d_{1q}$, we see that Figure 10.6 translates into β mapping the initial segment "$(c_{11} \cdots c_{1p}$" onto "$d_{1q} \cdots d_{1q_1}]$". (*Case* II) If $x = c_{1m}$ where $m > 1$, we "backtrack" (Figure 10.7). That is, we may view $\alpha \circ \beta = \beta \circ \alpha$ in terms of the diagrams in Figure 10.2 and thereby obtain

$$\begin{array}{ccc} c_{1,m-1} & \xrightarrow{\alpha} & c_{1m} = x \\ \beta\big\downarrow & & \big\downarrow\beta \\ c_{1,m-1}\beta & \xrightarrow[\alpha]{} & c_{1m}\beta = x\beta. \end{array}$$

Figure 10.7. Mapping initial segments onto terminal segments.

This backtrack argument shows that when β is defined at c_{1m}, then β must also be defined at $c_{1,m-1}$. Ultimately, then, we must have β mapping the initial segment "$(c_{11} \cdots c_{1p}$" onto the terminal segment "$d_{1q} \cdots d_{1q_1}]$". In conclusion, β maps an initial segment of $(c_{11} \cdots c_{1p}]$ onto a terminal segment of $(d_{11} \cdots d_{1q_1}]$, and, since β is one-one, an induction argument on the number u of proper paths shows that (2) is true.

Conversely, suppose β satisfies both properties (1) and (2). First, $\mathbf{d}(\alpha \circ \beta) = \mathbf{d}(\beta \circ \alpha)$: If $x \in \mathbf{d}(\alpha \circ \beta)$, then either (i) $x\alpha = a_{ij}$ or (ii) $x\alpha = c_{ij}$ for

$j > i$. If (i), then from (1), $x\beta \in \mathbf{d}\alpha$; and if (ii), then $x = c_{i,j-1}$ and, from (2), $x\beta \in \mathbf{d}\alpha$. So in either case, $x \in \mathbf{d}(\beta \circ \alpha)$. To see the reverse inclusion, let $x \in \mathbf{d}(\beta \circ \alpha)$. Then either (iii) $x\beta = b_{ij}$ or (iv) $x\beta = d_{ij}$. If (iii), then from (1), $x\alpha \in \mathbf{d}\beta$; and if (iv), then $j < q_i$ since $x \in \mathbf{d}(\beta\alpha)$, and, from (2), $x\alpha \in \mathbf{d}\beta$. And so $\mathbf{d}(\alpha \circ \beta) = \mathbf{d}(\beta \circ \alpha)$. Second, for $x \in \mathbf{d}(\alpha \circ \beta)$, we show $x(\alpha \circ \beta) = x(\beta \circ \alpha)$: If $x = a_{ij}$, then, with the obvious meaning of $a_{i,j+1}$,

$$(a_{ij})\alpha \circ \beta = a_{i,j+1}\beta = b_{i,j+1} = b_{ij}\alpha = (a_{ij})\beta \circ \alpha.$$

If $x = c_{ij}$, then $j < p_i$ and

$$(c_{ij})\alpha \circ \beta = c_{i,j+1}\beta = d_{i,m} = d_{i,m-1}\alpha = (c_{i,j})\beta \circ \alpha.$$

Thus, $\alpha \circ \beta = \beta \circ \alpha$, which finishes the proof. □

10.8 Corollary (Commuting Permutations)

Let $\alpha, \beta \in S_n$, and let

$$\alpha = (a_{11} \cdots a_{1r_1})(a_{21} \cdots a_{2r_2}) \cdots (a_{v1} \cdots a_{vr_v})$$

be the disjoint cycle decomposition of α. Then $\alpha \circ \beta = \beta \circ \alpha$ if and only if the disjoint cycle decomposition of α may be written as

$$\alpha = (b_{11} \cdots b_{1r_1})(b_{21} \cdots b_{2r_2}) \cdots (b_{v1} \cdots b_{vr_v})$$

where $a_{ij}\beta = b_{ij}$ for all indices i and j.

10.9 Corollary

Let $\alpha \in S_n$ have path decomposition $\alpha = (a_{11} \cdots a_{1\ell}) \cdots (a_{m1} \cdots a_{m\ell})$, i.e., let α be a regular permutation. Then the order of the centralizer $C(\alpha) = \{\,\beta \in S_n \,|\, \alpha \circ \beta = \beta \circ \alpha\,\}$ is $\ell^m m!$.

Proof. By 10.8, a permutation $\beta \in C(\alpha)$ if and only if β induces a matching of the m ℓ-circuits, and then preserves the cyclic ordering for each matched pair of ℓ-circuits. Since there are $m!$ ways to match the m circuits and, independent of each such matching, there are ℓ ways to map a particular ℓ-circuit onto an ℓ-circuit while preserving the cyclic ordering, it follows that there are $\ell^m m!$ members of $C(\alpha)$. □

10.10 Corollary

Let S be a finite inverse semigroup. Then S is a regular semigroup and the idempotents of S commute.

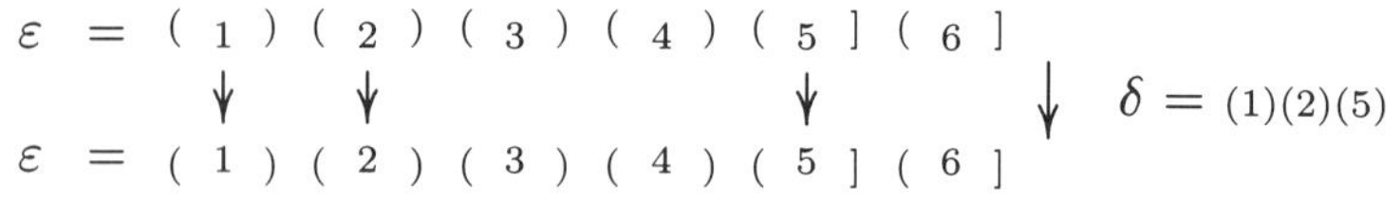

$$\begin{array}{l} \varepsilon = (\,1\,)(\,2\,)(\,3\,)(\,4\,)(\,5\,]\,(\,6\,] \\ \quad\ \ \downarrow\quad\ \downarrow\qquad\qquad\qquad\ \downarrow\qquad\quad \downarrow\ \ \delta = (1)(2)(5) \\ \varepsilon = (\,1\,)(\,2\,)(\,3\,)(\,4\,)(\,5\,]\,(\,6\,] \end{array}$$

Figure 10.11. Picturing that idempotents commute.

Proof. Clearly S is regular. So first, use (1) of 5.1 and apply 10.1 to show that idempotents in C_n commute (Figure 10.11). The result then follows from the Wagner-Preston Theorem. □

§11 Comments

Corollary 10.8 seems to be part of group theory folklore — $\alpha \circ \beta = \beta \circ \alpha \Leftrightarrow \beta^{-1} \circ \alpha \circ \beta = \alpha$, and the "first principle" in §32 of C. Jordan's 1870 text [1] is that the cycles in the disjoint cycle decomposition of $\beta^{-1} \circ \alpha \circ \beta$ are obtained from those of α by replacing each letter a_{ij} with $a_{ij}\beta$.

The C_n counterpart to 10.8, namely 10.1, was introduced by the author [2] and is fundamental to our study of centralizers (Chapters 5 and 6). Recently, using the path decomposition of partial transformations (Chapter 11), Theorem 10.1 was extended to a characterization of $\alpha \circ \beta = \beta \circ \alpha$ where $\alpha, \beta \in PT_n$, the semigroup of partial (not necessarily one-one) transformations $N \to N$ (Konieczny and Lipscomb [1]). An alternative proof of 10.9 appears in §170 of Burnside's text [1], and 10.10 is the finite case of Liber's [1] 1953 classic result.

Within the general theory of semigroups, studies of commutativity, in various guises, are indeed both plentiful and diverse, but characterizations of commuting charts prior to the 1988 discovery of 10.1 seem to be nonexistent. For permuting full (not necessarily one-one) transformations, see Ehrenfeucht and Harju and Rozenberg [1].

CHAPTER 4

Centralizers of Permutations

The study of centralizers of permutations, whether relative to the symmetric group S_n or the symmetric inverse semigroup C_n, reduces to the case of regular permutations. The case of regular permutations, in turn, involves wreath products — the cycle structure of a regular permutation $\alpha \in S_n$ determines a wreath product W, and it is through the choice of W for the C_n case that the S_n and C_n theories are unified. In fact, while this W is not a monoid, it is of course a semigroup, and it includes (as a subgroup) the wreath product W' used in the S_n case.[1]

With $W' \subset W$, we extend the classical isomorphism from $\{\, \beta \in S_n \mid \beta \circ \alpha = \alpha \circ \beta \,\}$ onto W' to a faithful representation of the centralizer $C(\alpha) = \{\beta \in C_n \mid \alpha \circ \beta = \beta \circ \alpha\}$ into W.

The approach is based on 10.1, the "Commuting Charts Theorem," and runs parallel to the group case. We also provide a formula for calculating orders of centralizers in C_n.

§12 Examples of Centralizers

Because of the reduction — permutations $\rightarrow$ regular permutations, it is instructive to study simple examples of centralizers of regular permutations.

Let $\alpha = (12)(34) \in S_4$. Then the elements of the S_4-centralizer $\{\, \beta \in S_4 \mid \beta \circ \alpha = \alpha \circ \beta\}$ are listed in the right column of Table 12.1, with the middle column serving as a memory aid and illustrating applications of Corollary 10.8. (Recall Figure 9.1.) In this particularly simple case, $\beta \circ \alpha = \alpha \circ \beta$ means that β induces a permutation defined on $\{(12), (34)\}$.

It is well known that the S_4-centralizer of α is isomorphic to $W' = Z_2$ wr S_2, where S_2 encodes all permutations defined on $\{(12), (34)\}$, and where $Z_2 = \{0, 1\}$, denoting the cyclic group of order two, accounts for the existence of two order-preserving maps between two 2-cycles.

Turning to the C_4-centralizer $C(\alpha)$, we see its 17 charts listed in the right column of Table 12.2. Again, the middle column serves as a memory aid, this time illustrating applications of Theorem 10.1. (Recall Figure 9.2.) The entries in the left column comprise the appropriate members of the wreath

[1]The wreath product W' is defined in §74; or it may be defined by appropriately modifying the definition of W that is given in §14, i.e., substitute S_m for C_m, and Z_ℓ for Z_ℓ^z.

Table 12.1. The S_4-centralizer of $\alpha = (12)(34)$.

Z_2 wr S_2	mnemonic	$C(\alpha)$
1. $(00, (1)(2))$	$\binom{12}{12}\binom{34}{34}$	1. $(1)(2)(3)(4)$
2. $(10, (1)(2))$	$\binom{12}{21}\binom{34}{34}$	2. $(12)(3)(4)$
3. $(01, (1)(2))$	$\binom{12}{12}\binom{34}{43}$	3. $(1)(2)(34)$
4. $(11, (1)(2))$	$\binom{12}{21}\binom{34}{43}$	4. $(12)(34)$
5. $(00, (12))$	$\binom{12}{34}\binom{34}{12}$	5. $(13)(24)$
6. $(01, (12))$	$\binom{12}{34}\binom{34}{21}$	6. (3241)
7. $(10, (12))$	$\binom{12}{43}\binom{34}{12}$	7. (3142)
8. $(11, (12))$	$\binom{12}{43}\binom{34}{21}$	8. $(32)(41)$

Table 12.2. The C_4-centralizer of $\alpha = (12)(34)$.

$C(\alpha)$ in Z_2^z wr C_2	mnemonic	$C(\alpha)$
1. $(00, (1)(2))$	$\binom{12}{12}\binom{34}{34}$	1. $(1)(2)(3)(4)$
2. $(10, (1)(2))$	$\binom{12}{21}\binom{34}{34}$	2. $(12)(3)(4)$
3. $(01, (1)(2))$	$\binom{12}{12}\binom{34}{43}$	3. $(1)(2)(34)$
4. $(11, (1)(2))$	$\binom{12}{21}\binom{34}{43}$	4. $(12)(34)$
5. $(00, (12))$	$\binom{12}{34}\binom{34}{12}$	5. $(13)(24)$
6. $(01, (12))$	$\binom{12}{34}\binom{34}{21}$	6. (3241)
7. $(10, (12))$	$\binom{12}{43}\binom{34}{12}$	7. (3142)
8. $(11, (12))$	$\binom{12}{43}\binom{34}{21}$	8. $(32)(41)$
9. $(0z, (1)(2])$	$\binom{12}{12}\binom{}{34}$	9. $(1)(2)$
10. $(1z, (1)(2])$	$\binom{12}{21}\binom{}{34}$	10. (12)
11. $(z0, (1](2))$	$\binom{}{12}\binom{34}{34}$	11. $(3)(4)$
12. $(z1, (1](2))$	$\binom{}{12}\binom{34}{43}$	12. (34)
13. $(0z, (12])$	$\binom{12}{34}\binom{}{12}$	13. $(13](24]$
14. $(1z, (12])$	$\binom{12}{43}\binom{}{12}$	14. $(14](23]$
15. $(z0, (21])$	$\binom{}{34}\binom{34}{12}$	15. $(31](42]$
16. $(z1, (21])$	$\binom{}{34}\binom{34}{21}$	16. $(32](41]$
17. $(zz, (1](2])$	$\binom{}{12}\binom{}{34}$	17. $(1]$

product $W = Z_2^z \text{ wr } C_2$, where C_2 encodes the charts defined on $\{(12),(34)\}$, and where Z_2^z, denoting Z_2 with a zero z adjoined, accounts not only for the existence of two order-preserving maps between two 2-cycles, but also for the possibility of not having a map between two 2-cycles.

Multiplication in $Z_2^z \text{ wr } C_2$ is defined so that the bijection between the entries in the left column with those on the right is an isomorphism.

§13 Centralizers as Direct Products

Let S be a semigroup and suppose $a \in S$. As usual, the centralizer of a in S is the set $C_S(a) = \{x \in S \mid ax = xa\}$ of elements that commute with a. It is clearly a subsemigroup of S and, whenever appropriate, we shall simply denote it as $C(a)$.

When S is the symmetric inverse semigroup C_n and α is a permutation in S_n, then the most useful characterization of $\beta \in C(\alpha)$ is 10.1, which involves the path structure of α. The following corollary is a special case of 10.1.

13.1 Corollary

Let $\alpha \in S_n$, let $\beta \in C_n$, and let α have path decomposition

$$\alpha = (a_{11}a_{12}\cdots a_{1\ell_1})(a_{21}a_{22}\cdots a_{2\ell_2})\cdots(a_{m1}a_{m2}\cdots a_{m\ell_m}).$$

Then $\alpha \circ \beta = \beta \circ \alpha$ if and only if α has path decomposition

$$(b_{11}b_{12}\cdots b_{1\ell_1})(b_{21}b_{22}\cdots b_{2\ell_2})\cdots(b_{m1}b_{m2}\cdots b_{m\ell_m})$$

where if $a_{ij} \in \mathbf{d}\beta$, then $a_{i1},\ldots,a_{i\ell_i} \in \mathbf{d}\beta$ and, for every $j = 1,\ldots,\ell_i$, we have $a_{ij}\beta = b_{ij}$.

It follows that we may construct a chart β that commutes with α by following a two-step process. First, permute some (possibly none) of the α circuits, i.e., "operate" on the first subscript i of the a_{ij}. Second, for corresponding (matched by our first step) α circuits, choose a bijection that preserves the α-induced cyclic ordering, i.e., "operate" on the second subscript j of the a_{ij}. Such two-step processes generally yield to wreath products; and, in this case, our wreath product connects with $C(\alpha)$ by way of regular permutations.

To see the connection, suppose that $\alpha \in S_n$, that $\beta \in C_n$, and that $\alpha \circ \beta = \beta \circ \alpha$. For this α, let K be the set of positive integers k for which there exists a k-circuit in the path decomposition of α, and, for each $k \in K$, suppose N_k denotes the set of letters a_{ij} with $\ell_i = k$. Then β leaves N_k invariant, i.e., $N_k\beta \subset N_k$. In short, every $\beta \in C(\alpha)$ leaves each N_k invariant.

13.2 Proposition

A chart $\beta \in C_n$ commutes with $\alpha \in S_n$ if and only if for each $k \in K$, we have $N_k\beta \subset N_k$ and the restrictions β_k and α_k, of β and α to N_k, commute.

Since $N_k \cap N_{k'} = \emptyset$ for $k \neq k'$, we may view $C(\alpha)$ as a direct product

$$C(\alpha) = \times_K C_k(\alpha_k)$$

where $C_k(\alpha_k)$ is the centralizer (relative to C_{N_k}) of $\alpha_k \in S_{N_k}$. Moreover, since each permutation α_k is regular, $C(\alpha)$ is a direct product of centralizers of regular permutations.

§14 Centralizers and Wreath Products

We assume that $\alpha \in S_n$ is regular. That is, we now assume that the path decomposition of α has the following form:

$$\alpha = (a_{10}a_{11}\cdots a_{1,\ell-1})(a_{20}\cdots a_{2,\ell-1})\cdots(a_{m0}\cdots a_{m,\ell-1}) \in S_n.$$

For such an α, we may view $C(\alpha)$ as a submonoid of a wreath product. To that end, note that $n = m\ell$, define $P = \{1, 2, \ldots, m\}$, and consider the symmetric inverse semigroup C_m. These choices essentially allow for permuting some (possibly none) of the first indices on the letters appearing in the ℓ-circuits of α. For encoding shifts of second indices, select the semigroup $S = Z_\ell^z$, the cyclic group Z_ℓ with zero z adjoined. Then, with S^P denoting the set of all mappings $P \to Z_\ell^z$, define Z_ℓ^z wr C_m as the set $W = S^P \times C_m$ with multiplication $W \times W \to W$ given by

$$(f,t)(g,u) = (fg^t, tu) \quad \text{where} \quad p(fg^t) = \begin{cases} pf + (pt)g \in S & \text{if } p \in \mathbf{d}t, \\ z \in S & \text{if } p \notin \mathbf{d}t. \end{cases}$$

It is straightforward to show that this binary operation is associative, making W a semigroup. However, W is not a monoid: Consider $(f,t) \in W$ such that $p \in \mathbf{d}f - \mathbf{d}t = P - \mathbf{d}t$ and $pf \neq z$. Then $(f,t)(g,u) = (f,t)$ implies $fg^t = f$, which is contradicted by $p(fg^t) = z \neq pf$. But W does have a zero $(f_z, (1])$, where $f_z : P \to Z_\ell^z$ is the constant map whose value is z. We also note that Z_ℓ wr S_m is a subgroup of Z_ℓ^z wr C_m.

14.1 Theorem

Let $\alpha \in S_n$ be a regular permutation whose path decomposition contains m ℓ-circuits. Then the centralizer $C(\alpha)$ in C_n is isomorphic to a submonoid of the semigroup Z_ℓ^z wr C_m.

Proof. For $W = Z_\ell^z$ wr C_m, we define a monomorphism $\theta : C(\alpha) \to W$. To that end, let $\beta \in C(\alpha)$ and define a chart $\phi_\beta \in C_m$ and a function $f_\beta : P \to Z_\ell^z$ as follows: For those $i \in P$ such that $a_{i0} \in \mathbf{d}\beta$, we have $a_{i0}\beta = a_{jk}$, which allows us to define

$$j = i\phi_\beta \quad \text{and} \quad k = if_\beta.$$

And for those $i \in P$ such that $a_{i0} \notin \mathbf{d}\beta$, we shall agree that

$$i\phi_\beta \text{ is undefined}, \quad if_\beta = z.$$

It follows from 10.1 that both ϕ_β and f_β are well defined, and that ϕ_β is one-one. We let $\theta : C(\alpha) \to W$ be the map $\beta \mapsto (f_\beta, \phi_\beta)$. To see that θ is a homomorphism, we begin with $i \in \mathbf{d}\phi_\beta$. Then, since β preserves the cyclic ordering of $(a_{i0}, \ldots, a_{i,\ell-1})$,

$$a_{ir}\beta = (a_{i0}\alpha^r)\,\beta = a_{j,k+r}$$

where $j = i\phi_\beta$ and $k = if_\beta$. Moreover, for $\gamma \in C(\alpha)$ such that $a_{j0} \in \mathbf{d}\gamma$, we calculate that

$$(a_{i0})\beta\gamma = (a_{jk})\gamma = a_{u,v+k}$$

where $u = j\phi_\gamma$ and $v = jf_\gamma$. It follows, for $i \in \mathbf{d}\phi_{\beta\gamma}$, that

$$i\phi_{\beta\gamma} = i\phi_\beta\phi_\gamma \quad \text{and} \quad if_{\beta\gamma} = jf_\gamma + if_\beta = i\phi_\beta f_\gamma + if_\beta.$$

Thus, since $\mathbf{d}\phi_{\beta\gamma} = \mathbf{d}(\phi_\beta\phi_\gamma)$, as charts

$$\phi_{\beta\gamma} = \phi_\beta\phi_\gamma.$$

Moreover, since

$$if_{\beta\gamma} = if_\beta + i\phi_\beta f_\gamma$$

whenever $i \in \mathbf{d}\phi_{\beta\gamma}$, we only need to consider the case where $i \notin \mathbf{d}\phi_{\beta\gamma}$. In this case, $if_{\beta\gamma} = z$, and either if_β or $i\phi_\beta f_\gamma$ equals z. In summary, therefore,

$$f_{\beta\gamma} = f_\beta + \phi_\beta f_\gamma,$$

showing that θ is a morphism. That θ is a monomorphism follows from its definition and 10.1. □

14.2 Corollary

Let $\alpha \in S_n$ be a regular permutation whose disjoint cycle decomposition has m cycles each of length ℓ. Then $C(\alpha)$, relative to S_n, is isomorphic to the subgroup Z_ℓ wr S_m of Z_ℓ^z wr C_m and has order $\ell^m m!$.

Proof. Using the morphism θ constructed in the proof of 14.1, we observe that $\beta \in C_{C_n}(\alpha)$ is a permutation if and only if $z \notin \mathbf{r}f_\beta$ and $\phi_\beta \in S_m$. It follows that θ maps $C_{S_n}(\alpha)$ onto Z_ℓ wr S_m, which is a subgroup of Z_ℓ^z wr C_m. And since this restriction of θ is an isomorphism onto Z_ℓ wr S_m, the order of Z_ℓ wr S_m, namely $\ell^m m!$, is the order of the $C_{S_n}(\alpha)$. □

14.3 Corollary

If $\alpha \in S_n$ is a regular permutation whose path decomposition has precisely n_k k-circuits, then the order of $C(\alpha)$, relative to C_n, is

$$\textstyle\sum_{j=0}^{n_k} \binom{n_k}{j}^2 j!k^j = m_k,$$

and if $\alpha \in S_n$ is an arbitrary permutation whose path decomposition has precisely n_k k-circuits, where $n = n_1 + 2n_2 + \cdots + nn_n$, then the order of $C_{C_n}(\alpha)$ is

$$\Pi_k m_k.$$

Proof. It follows from the direct product representation of $C(\alpha)$ that the number $|C(\alpha)| = \Pi_k |C_k(\alpha_k)|$ where $C_k(\alpha_k)$ is the centralizer (relative to C_{N_k}) of $\alpha_k \in S_{N_k}$. So the only problem is the calculation of each $|C_k(\alpha_k)|$. Indeed, for α with n_k k-circuits and $n = n_k k$, we see that to choose β that commutes with α_k is to first choose j, $0 \le j \le n_k$, of the n_k circuits for the domain of β and then choose j of the n_k circuits for the range of β. Clearly, there are $j!$ ways to map the domain j-set of cycles one-one onto the range j-set of circuits. Lastly, for each circuit in the domain and its corresponding circuit in the range, there are k ways to map the former onto the latter so that the cyclic order is preserved. □

To illustrate 14.3, let us return to our example of $C(\alpha)$ where $\alpha = (12)(34)$. In this case $\alpha \in S_4$ is a regular permutation with n_k k-circuits where $n_k = k = 2$. We therefore calculate that

$$\textstyle\sum_{j=0}^{n_k} \binom{n_k}{j}^2 j!k^j = \sum_{j=0}^{2} \binom{2}{j} j!2^j = 1 + 8 + 8 = 17,$$

which agrees with Table 12.2.

§15 Comments

Using the isomorphism in 14.2 to prove that $|C(\alpha)| = \ell^m m!$ is not as direct as the combinatorial approach in 10.9. However, with the isomorphism $C(\alpha) \to (Z_\ell \text{ wr } S_m)$ followed by the projection $(Z_\ell \text{ wr } S_m) \to S_m$ morphism, we see that $C(\alpha)$ must contain a normal subgroup H of order ℓ^m and that

$C(\alpha)/H$ has order $m!$. The group H is described in §170 of Burnside's 1911 text; the facts are as follows:

If H is the subgroup of $C(\alpha)$ generated by the cycles

$$(a_{11} \cdots a_{1\ell}), \ldots, (a_{m1} \cdots a_{m\ell})$$

that appear in the disjoint cycle decomposition of the regular α, then H is normal in $C(\alpha)$ and has order ℓ^m. Moreover, $C(\alpha)/H$ is isomorphic to the subgroup Q of $C(\alpha)$ generated by the two elements

$$\begin{aligned}&(a_{11}a_{21} \cdots a_{m1})(a_{12}a_{22} \cdots a_{m2}) \cdots (a_{1\ell}a_{2\ell} \cdots a_{m\ell}) \quad \text{and} \\ &(a_{11}a_{21})(a_{12}a_{22}) \cdots (a_{1\ell}a_{2\ell}).\end{aligned}$$

It follows that Q must have order $m!$, and, since $H \cap Q$ contains only the identity, that $C(\alpha)$ is a semidirect product of H by Q.

Corollary 14.2 seems to be standard, e.g., in Suzuki's text [1, page 296], we find a statement similar to 14.2 but involving a different "wreath product." By extending the wreath product used by Suzuki, the author [2] managed to prove Corollary 14.3 and a statement similar to 14.1. Comparing those "wreath products" with the one used here, the difference turns on the definition of f_β (proof of 14.1). The f_β defined in the proof of 14.1 not only seems more natural, but the corresponding wreath product is of the same ilk as the one used to study centralizers of arbitrary charts (Chapter 5).

These wreath products are fairly standard, e.g., in addition to Suzuki [1], see Chapter 4 of Lallement [1], and for wreath products of partial transformation semigroups, see Chapter 1 of Eilenberg [2]. The "n-fold wreath products" serve to define *decompositions of transformation semigroups*, which, in turn, are basic to the important Krohn-Rhodes Decomposition Theorem (Eilenberg [2, Chapter 2]).

Studies of centralizers outside of group theory but within the general theory of transformation semigroups do exist, e.g., see Ehrenfeucht and Harju and Rozenberg [1] and Higgins [2], which are discussed in §61 of Chapter 12. But results on centralizers of permutations in C_n prior to the 1988 discovery of 14.1 seem to be nonexistent.

CHAPTER 5

Centralizers of Charts

Centralizers of permutations were studied in Chapter 4, where the S_n case appeared as a corollary within the more general C_n theory. In this chapter, we study centralizers $C(\alpha)$ of charts $\alpha \in C_n - S_n$; the approach involves a quotient of a wreath product.

For any chart $\alpha = \gamma\eta \in C_n$, which has both a permutation part γ and a nilpotent part η, it turns out that $C(\alpha)$ is isomorphic to the direct product $C(\gamma) \times C(\eta)$. This isomorphism reduces our study of $C(\alpha)$ to separate studies of $C(\gamma)$ and $C(\eta)$: Since the order and faithful representation of $C(\gamma)$ are given in Chapter 4, our focus here is on $C(\eta)$. First, we imbed $C(\eta)$ in a quotient of a wreath product. Then we show that like $C(\gamma)$, whose order depends only on the lengths and number of circuits appearing in the path decomposition of γ, the order of $C(\eta)$ depends only on the lengths and number of proper paths appearing in the path decomposition of η.

§16 General Case

If a chart $\alpha = \gamma\eta \in C_n$ has permutation part γ and nilpotent part η, then we may view the centralizer $C(\alpha)$ as a direct product

$$C(\alpha) \simeq C(\gamma) \times C(\eta). \tag{16.1}$$

The existence of an isomorphism (16.1) follows from 10.1: Any member of $C(\alpha)$, in effect, matches circuits of α only with circuits of α, and proper paths of α only with proper paths of α. Therefore, letting $(\mathbf{d}\gamma)' = \{1, 2, \ldots, n\} - \mathbf{d}\gamma$, we see that if $\beta \in C(\alpha)$, then β is the join of $\beta|_{\mathbf{d}\gamma}$ and $\beta|_{(d\gamma)'}$. From this observation, and agreeing that $C(\gamma)$ is relative to the symmetric inverse semigroup $C_{\mathbf{d}\gamma}$ and that $C(\eta)$ is relative to $C_{(\mathbf{d}\gamma)'}$, it follows that the map $C(\alpha) \to C(\gamma) \times C(\eta)$ given by

$$\beta \mapsto (\beta|_{\mathbf{d}\gamma}, \beta|_{(\mathbf{d}\gamma)'})$$

is an isomorphism. In other words, to know the centralizers of all permutations and all nilpotents is to know the centralizers of all charts.

§17 Nilpotent Case: An Example

Unlike the permutation case, where the study reduces to regular permutations, the centralizer theory of nilpotents does not reduce to "regular nilpotents." Nevertheless, it is instructive to study a simple example where the nilpotent is regular.

Suppose $\eta = (12](34] \in C_4$. Then the 17 elements of the C_4-centralizer $C(\eta)$ are listed in the right column of Table 17.1. The middle column serves as a memory aid, illustrating applications of 10.1. (Recall Figure 9.2.) In effect, $\beta \circ \eta = \eta \circ \beta$ means that β, while preserving the η-induced linear ordering and determining a chart defined on $\{(12], (34]\}$, is a map of some initial segments of $(12]$ or $(34]$ onto some terminal segments of $(12]$ or $(34]$.

Table 17.1. The C_4-centralizer of $\eta = (12](34]$.

$(f, t) \in S$ wr C_2	mnemonic	$C(\eta)$
1. $(11, (1)(2))$	$\binom{12}{12}\binom{34}{34}$	1. $(1)(2)(3)(4)$
2. $(\mu 1, (1)(2))$	$\binom{12}{2}\binom{34}{34}$	2. $(12](3)(4)$
3. $(1\mu, (1)(2))$	$\binom{12}{12}\binom{34}{4}$	3. $(1)(2)(34]$
4. $(\mu\mu, (1)(2))$	$\binom{12}{2}\binom{34}{4}$	4. $(12](34]$
5. $(11, (12))$	$\binom{12}{34}\binom{34}{12}$	5. $(13)(24)$
6. $(1\mu, (12))$	$\binom{12}{34}\binom{34}{2}$	6. $(1324]$
7. $(\mu 1, (12))$	$\binom{12}{4}\binom{34}{12}$	7. $(3142]$
8. $(\mu\mu, (12))$	$\binom{12}{4}\binom{34}{2}$	8. $(32](41]$
9. $(1z, (1)(2])$	$\binom{12}{12}\binom{}{34}$	9. $(1)(2)$
10. $(\mu z, (1)(2])$	$\binom{12}{2}\binom{}{34}$	10. $(12]$
11. $(z1, (1](2))$	$\binom{}{12}\binom{34}{34}$	11. $(3)(4)$
12. $(z\mu, (1](2))$	$\binom{}{12}\binom{34}{4}$	12. $(34]$
13. $(1z, (12])$	$\binom{12}{34}\binom{}{12}$	13. $(13](24]$
14. $(\mu z, (12])$	$\binom{12}{4}\binom{}{12}$	14. $(14]$
15. $(z1, (21])$	$\binom{}{34}\binom{34}{12}$	15. $(31](42]$
16. $(z\mu, (21])$	$\binom{}{34}\binom{34}{2}$	16. $(32]$
17. $(zz, (1](2])$	$\binom{}{12}\binom{}{34}$	17. $(1]$

The entries in the left column of Table 17.1 comprise appropriate members of the wreath product $W = S$ wr C_2. In this case, using z to denote the zero of C_2, we have $S = \{1, \mu, z\}$, where $\{\mu, z\} = \langle \mu \rangle$ is the cyclic subsemigroup of C_2 generated by $\mu = (12]$ and $1 = (1)(2) \in C_2$ is the identity of C_2.

At this point, it is instructive to compare Table 17.1 with Table 12.2: In the permutation case (Table 12.2), the injection $C(\alpha) \to Z_2^z \text{ wr } C_2$ that matches the entries in the right column with those in the left is an imbedding. In our nilpotent case, however, the corresponding injection $C(\eta) \to W$ is not an imbedding — the square $(12](34] \circ (12](34]$ of the chart numbered "4" in the right column is the zero chart $(1]$, which is numbered "17," while, given that the multiplication in W induces component multiplication in the second component, $(\mu\mu, (1)(2))^2 \neq (zz, (1](2])$. This example is typical in that the defect is always among "second components." An adjustment will be made in §19, where we define a quotient map $W \to \mathcal{W}$ that identifies certain "second components," and in §20, where we use the quotient map to define an imbedding $C(\eta) \to W \to \mathcal{W}$.

In passing, we note that even though this example is rather simple, the choice of S is representative, i.e., S is generated by a proper path and certain idempotents, which illustrates the key aspects of the general theory.

§18 Nilpotent Case: The Wreath Product

Suppose $\eta \in C_n$ is nilpotent with path decomposition

$$\eta = (a_{10}a_{11}\cdots a_{1r_1}](a_{20}\cdots a_{2r_2}]\cdots(a_{m0}\cdots a_{mr_m}] \in C_n. \tag{18.1}$$

For such an η, our goal is to faithfully represent $C(\eta)$ in a quotient $\mathcal{W}$ of a wreath product W. To reach our goal, we begin with a few definitions. First, we encode the permuting of some (possibly none) of the first indices on the letters appearing in (18.1) by letting

$$P = \{1, 2, \ldots, m\},$$

and considering the symmetric inverse semigroup C_m. Second, we begin to encode mappings of initial segments onto terminal segments by letting $\ell = \max\{\, r_i + 1 \,|\, i = 1, \ldots, m\,\}$ be the maximum of the lengths of the proper paths appearing in (18.1) and defining $Q = \{0, 1, \ldots, \ell - 1\}$. Then, letting μ and ε_j, $0 \le j \le \ell - 1$, be the charts given by

$$\mu = (0, 1, \ldots, \ell - 1] \quad \text{and} \quad \varepsilon_j = (0)(1)\cdots(j)(j+1]\cdots(\ell-1],$$

we define S via

$$S = \{\, \mu^k \varepsilon_j \mid 0 \le j \le \ell - 1;\ k = 0, 1, \ldots \ell\} \subset C_Q,$$

where we shall understand that $\mu^0 = \varepsilon_{\ell-1} = (0)\cdots(\ell-1) = 1_Q$ is the identity in C_Q and $z = \mu^\ell$ is the zero in C_Q. Thus, $1_Q, z \in S$ and the semigroup (monoid) S encodes *initial-segment shifts* by restricting the members of the cyclic semigroup $\langle\mu\rangle$ and 1_Q to the "initial segments of μ." Indeed, the members of S "shift initial segments" of μ by $k = 0, 1, ..., \ell - 1$ "steps to the right." (As will become obvious from the proofs below, we need the "ε_j" on the right of $\mu^k \varepsilon_j$ because the decomposition of η may contain a proper path whose length is less than ℓ.)

Using S^P to denote the set of functions $P \to S$, we define the set $W = S^P \times C_m$ with the multiplication $W \times W \to W$ given by

$$(f,t)(g,u) = (fg^t, tu) \quad \text{where } p(fg^t) = \begin{cases} (pf)((pt)g) \in S & \text{if } p \in \mathbf{d}t, \\ z \in S & \text{if } p \notin \mathbf{d}t. \end{cases}$$

It is straightforward to show that this binary operation is associative, making W a semigroup. With the arguments given in §14, it is clear that W is not a monoid but that W has a zero.

To reflect on the meaning of $(f,t) \in W$ in the context of 10.1, first think of $t \in C_m$ as matching some of the proper paths appearing in the join representation of η — those indexed by $\mathbf{d}t$ with those indexed by $\mathbf{r}t$. Second, for each $p \in \mathbf{d}t$, think of $pf = s \in S$ as specifying a map of an initial segment of the pth proper path onto a terminal segment of the (pt)th proper path. For example, consider the decomposition of the nilpotent

$$\eta = (i_0 i_1 i_2 i_3 i_4 i_5](j_0 j_1 j_2] \in C_9.$$

Then, because there are two proper paths, $P = \{1, 2\}$, and because the maximum length path has length six, $\mu = (012345]$. In this case, suppose $t = (12] \in C_2$ and $(1, \mu^2 \varepsilon_2) \in f \in S^P$. Then $(f,t) \in W$ and t matches the first path with the second, i.e., matches the proper 6-path $(i_0, i_1, i_2, i_3, i_4, i_5]$ with the proper 3-path $(j_0, j_1, j_2]$, while the chart

$$1f = \mu^2 \varepsilon_2 = (024](135] \circ (0)(1)(2)(3](4](5] = (02](1](3](4](5]$$

maps the initial segment $\{i_0\}$ of the 6-path onto the terminal segment $\{j_2\}$ of the 3-path, illustrating the (commuting charts) Theorem 10.1.

§19 The Quotient Semigroup

The relation $\sim$ on W, given by $(f,t) \sim (f,u)$ whenever $pt = pu$ for all $p \in P$ such that $pf \neq z$, is not only an equivalence relation, it is a congruence (Proposition 19.1 below). We may therefore define $\mathcal{W}$ as the induced quotient semigroup, and use "$[(f,t)]$" to denote the equivalence class containing the representative $(f,t) \in W$.

19.1 Proposition

The equivalence $\sim$ is a congruence on W.

Proof. Let $(f,t) \sim (f,u)$, and let $(g,v) \sim (g,w)$. By definition of multiplication in W,

$$(f,t)(g,v) \sim (f,u)(g,w) \iff \big(h = fg^t = fg^u\big) \text{ and } \big(p(tv) = p(uw) \text{ when } ph \neq z\big). \tag{19.2}$$

But the first condition on the right-side of (19.2) holds: If $pf = z$, then, by definition of fg^t and fg^u, we have $pfg^t = pfg^u = z$. If $pf \neq z$, then, since $(f,t) \sim (f,u)$ implies $pt = pu$, we again have $pfg^t = pfg^u$. Thus, $fg^t = fg^u$. And the second condition on the right-side of (19.2) holds: If $ph \neq z$, then $pf \neq z$. Hence, $(f,t) \sim (f,u)$ implies

$$pt = pu. \tag{19.3}$$

With (19.3) and $ph \neq z$, it follows that

$$(pt)g = (pu)g \neq z. \tag{19.4}$$

From (19.4), using $(g,v) \sim (g,w)$ and the substitution (19.3), we have

$$(pt)v = (pt)w = (pu)w.$$

Thus, the right-side of (19.2) is true, completing the proof. □

§20 Imbedding $C(\eta)$ in the Quotient Semigroup

Throughout this section, we continue to assume that the nilpotent η has path decomposition (18.1) and that $\mathcal{W}$ is the quotient semigroup introduced in §19. We define a monomorphism $\theta : C(\eta) \to \mathcal{W}$. To that end, let $\rho \in C(\eta)$ and define a chart $t_\rho \in C_m$ and a function $f_\rho : P \to S$ as follows: For those $i \in P$ such that $a_{i0} \in \mathbf{d}\rho$, we have $a_{i0}\rho = a_{jq}$, which allows us to define

$$j = it_\rho \quad \text{and} \quad \mu^q \varepsilon_{r_j} = if_\rho. \tag{20.1}$$

For these values of i, recall that a_{jq} appears in $(a_{j0} \cdots a_{jr_j}]$ and, since we have $q \leq r_j < \ell$,

$$a_{i0} \in \mathbf{d}\rho \;\Rightarrow\; if_\rho \neq z. \tag{20.2}$$

For those $i \in P$ such that $a_{i0} \notin \mathbf{d}\rho$, we shall agree that

$$it_\rho \text{ is undefined} \quad \text{and} \quad if_\rho = z. \tag{20.3}$$

Moreover, for these values of i, we see that

$$i \notin \mathbf{d}(t_\rho) \Leftrightarrow if_\rho = z. \tag{20.4}$$

It is clear from (20.1), (20.3), and 10.1 that both $f_\rho \in S^P$ and $t_\rho \in C_m$. Consequently, we may define the mapping θ by

$$\rho\theta = \big[(f_\rho, t_\rho)\big]. \tag{20.5}$$

To prove that θ is a monomorphism, we need a technical lemma.

20.6 Lemma

Let the nilpotent $\eta \in C_n$ have path decomposition (18.1), and let $\theta : C(\eta) \to \mathcal{W}$ be given by (20.5). Then for $\rho, \tau \in C(\eta)$, we have, for each $p \in P$,

(*i*) $(pf_\rho)(pt_\rho)f_\tau = z \;\Leftrightarrow\; pf_{\rho\tau} = z$.
(*ii*) $pf_{\rho\tau} \neq z \;\Rightarrow\; \big(pf_{\rho\tau} = (pf_\rho)(pt_\rho)f_\tau \;\text{ and }\; pt_\rho t_\tau = pt_{\rho\tau}\big)$.

Proof. Proof of (*i*): ($\Leftarrow$) Suppose $pf_{\rho\tau} = z$. Then (20.2) shows that $a_{p0} \notin \mathbf{d}(\rho\tau)$. Hence, either $a_{p0} \notin \mathbf{d}\rho$, or $a_{p0}\rho = a_{jq}$ where $a_{jq} \notin \mathbf{d}\tau$. In the former case, by definition of f_ρ, we have $pf_\rho = z$ and then

$$(pf_\rho)(pt_\rho)f_\tau = z \cdot (pt_\rho)f_\tau = z. \tag{20.7}$$

In the latter case, again by definition of f_ρ, we have $pf_\rho = \mu^q \varepsilon_{r_j}$ and either (a) $a_{p0}\rho = a_{jq}$ with $a_{jq}, a_{j0} \notin \mathbf{d}\tau$, or (b) $a_{p0}\rho = a_{jq}$ with $a_{jq} \notin \mathbf{d}\tau$ but $a_{j0} \in \mathbf{d}\tau$. If (a), then $a_{j0} \notin \mathbf{d}\tau$ and (20.3) yield $jf_\tau = z$. But $j = pt_\rho$, since $a_{p0}\rho = a_{jq}$, and we conclude that $jf_\tau = (pt_\rho)f_\tau = z$. Thus, $(pf_\rho)(pt_\rho)f_\tau = \mu^q \varepsilon_{r_j} \cdot z = z$. If (b), then $a_{j0}\tau = a_{kh}$, implying $jf_\tau = \mu^h \varepsilon_{r_k}$. Whence,

$$(pf_\rho)(pt_\rho)f_\tau = \mu^q \varepsilon_{r_j} \cdot \mu^h \varepsilon_{r_k} = \mu^{q+h} \varepsilon_{r_j+h} \varepsilon_{r_k}, \tag{20.8}$$

where $\varepsilon_{r_j}\mu^h = \mu^h \varepsilon_{r_j+h}$ if we define $\varepsilon_{r_j+h} = \varepsilon_{\ell-1}$ when $r_j + h \geq \ell - 1$. Also, $a_{j0}\tau = a_{kh}$ shows that τ must map $a_{j0} \mapsto a_{kh}$, $a_{j1} \mapsto a_{k,1+h}$, $\ldots$, $a_{j,r_k-h} \mapsto a_{k,r_k}$. But because $a_{jq} \notin \mathbf{d}\tau$, we see that $q > r_k - h$, or $r_k < q+h$. The last inequality implies that the last expression on the right side of (20.8) must equal z. We therefore have $(pf_\rho)(pt_\rho)f_\tau = z$ when either (a) or (b) holds. This result, together with the case yielding (20.7), finishes the proof of the sufficiency part of (*i*). ($\Rightarrow$) So now let

$$(pf_\rho)(pt_\rho)f_\tau = z. \tag{20.9}$$

If $pf_\rho = z$, then $a_{p0} \notin \mathbf{d}\rho \Rightarrow a_{p0} \notin \mathbf{d}\rho\tau \Rightarrow pf_{\rho\tau} = z$. If $pf_\rho \neq z$, then $a_{p0}\rho = a_{jq}$, implying $pf_\rho = \mu^q \varepsilon_{r_j}$ where $q \leq r_j < \ell$. Substituting for pf_ρ in (20.9), we now have

$$\mu^q \varepsilon_{r_j} \cdot jf_\tau = z.$$

If $jf_\tau = z$, then $a_{j0} \notin \mathbf{d}\tau \Rightarrow a_{jq} \notin \mathbf{d}\tau \Rightarrow a_{p0} \notin \mathbf{d}\rho\tau \Rightarrow pf_{\rho\tau} = z$. We are therefore left with the case $jf_\tau \neq z$. So suppose $jf_\tau = \mu^h\varepsilon_{r_k}$, i.e., $a_{j0}\tau = a_{kh}$. Then, since $a_{p0}\rho = a_{jq}$, we have ($a_{p0} \in \mathbf{d}\rho\tau \Leftrightarrow q + h \leq r_k$). So (20.9) becomes

$$\mu^q\varepsilon_{r_j}\mu^h\varepsilon_{r_k} = \mu^{q+h}\varepsilon_{r_j+h}\varepsilon_{r_k} = z.$$

And, if $\mu^{q+h} = z$, then $q + h \geq \ell > r_k$ implies $a_{p0} \notin \mathbf{d}\rho\tau$, which, in turn, yields $pf_{\rho\tau} = z$. Thus, we are left with $\mu^{q+h} \neq z$: Then $q + h > \min\{r_j + h, r_k\} = r_k$ (the "=" following from $q \leq r_j$) and, we again conclude $a_{p0} \notin \mathbf{d}\rho\tau$, which shows, as before, that $pf_{\rho\tau} = z$. We now have $pf_{\rho\tau} = z$ in every possible case, finishing the proof of the necessity part of (i).
Proof of (ii): Suppose $pf_{\rho\tau} \neq z$. Using the notation in the preceding paragraph, we observe that

$$(pf_\rho)(pt_\rho)f_\tau = \mu^q\varepsilon_{r_j}\mu^h\varepsilon_{r_k} = \mu^{q+h}\varepsilon_{r_j+h}\varepsilon_{r_k},$$

where here we must have $r_k \leq r_j + h$, since otherwise, $r_j + h < r_k$ implies τ would not map an initial segment of $(a_{j0}\cdots a_{jr_j}]$ onto an a terminal segment of $(a_{k0}\cdots a_{kr_k}]$. Thus,

$$(pf_\rho)(pt_\rho)f_\tau = \mu^{q+h}\varepsilon_{r_j+h}\varepsilon_{r_k} = \mu^{q+h}\varepsilon_{r_k} = pf_{\rho\tau}.$$

Turning to the other equality in (ii), we observe that $pf_{\rho\tau} \neq z$ implies $a_{p0}(\rho\tau) = a_{kc}$, or

$$(a_{p0}\rho)\tau = (a_{jq})\tau = a_{k,q+h}, \qquad q + h = c \leq r_k,$$

where $a_{j0}\tau = a_{kh}$. But then $pt_{\rho\tau} = k = jt_\tau = (pt_\rho)t_\tau$. □

20.10 Theorem

Let $\eta \in C_n$ have path decomposition $(a_{10}a_{11}\cdots a_{1r_1}]\cdots(a_{m0}\cdots a_{mr_m}]$, and let $\theta : C(\eta) \to \mathcal{W}$ be given by $\rho \mapsto [(f_\rho, t_\rho)]$ where f_ρ and t_ρ are defined by (20.1) *and* (20.3). *Then θ is a monomorphism.*

Proof. First, we show that θ is injective: Suppose $\rho, \tau \in C(\eta)$ and $\rho\theta = \tau\theta$, i.e., $(f_\rho, t_\rho) \sim (f_\tau, t_\tau)$ or, $f_\rho = f_\tau = f$ and $pt_\rho = pt_\tau$ when $pf \neq z$. From (20.2), (20.4), and $f_\rho = f_\tau$,

$$p \notin \mathbf{d}(t_\rho) \;\Leftrightarrow\; pf_\rho = z \;\Leftrightarrow\; pf_\tau = z \;\Leftrightarrow\; p \notin \mathbf{d}(t_\tau).$$

Thus, according to 10.1, if $\rho \neq \tau$, then they differ at some $a_{io} \in \mathbf{d}\rho \cap \mathbf{d}\tau$. But if this were the case, then

$$a_{i0}\rho = a_{jq} \quad \text{and} \quad a_{i0}\tau = a_{jk},$$

where we must have $q \neq k$, since $j = it_\rho = it_\tau$ (because $if \neq z$). But $q \neq k$ and both $q, k \leq r_j < \ell$ imply

$$if_\rho = \mu^q \varepsilon_{r_j} \neq \mu^k \varepsilon_{r_j} = if_\tau,$$

which contradicts $f_\rho = f_\tau$. Consequently, $\rho = \tau$ and θ is injective.
Second, we show that θ is a homomorphism: Since we have $\rho \mapsto [(f_\rho, t_\rho)]$ and $\tau \mapsto [(f_\tau, t_\tau)]$ and $\rho\tau \mapsto [(f_{\rho\tau}, t_{\rho\tau})]$, it suffices to show $(f_\rho, t_\rho)(f_\tau, t_\tau) \sim (f_{\rho\tau}, t_{\rho\tau})$, i.e.,

$$\big((pf_\rho)(pt_\rho)f_\tau = pf_{\rho\tau}\big) \text{ and } \big(\text{when } pf_{\rho\tau} \neq z, \text{ then } t_\rho t_\tau = t_{\rho\tau}\big).$$

But (i) of 20.6 shows that $(pf_\rho)(pt_\rho)f_\tau = z \Leftrightarrow pf_{\rho\tau} = z$, and (ii) of 20.6 shows that if $pf_{\rho\tau} \neq z$, then $(pf_\rho)(pt_\rho)f_\tau = pf_{\rho\tau}$. Also, (ii) of 20.6 shows that (when $pf_{\rho\tau} \neq z$, then $t_\rho t_\tau = t_{\rho\tau}$). □

§21 Calculations

The right column of Table 17.1 contains a list of the members of the centralizer $C(\eta)$ of the nilpotent $\eta = (12](34]$. And in §17, we exhibited values where the injection $C(\eta) \to W$ does not preserve multiplication — η corresponds to $(\mu\mu, (1)(2))$, but η^2 does not correspond to $(\mu\mu, (1)(2))^2$. We also promised to define a quotient map $W \to \mathcal{W}$ to correct this defect, i.e., under the injection $C(\eta) \to W \to \mathcal{W}$, the square η^2 would correspond to $[(\mu\mu, (1)(2))]^2$.

Having now constructed, for any nilpotent η, a wreath product W_η and a quotient map $W_\eta \to \mathcal{W}_\eta$, we return to that example to test our claim. We begin by adopting the more general notation. In particular, we now have $\eta = (a_{10}a_{11}](a_{20}a_{21}]$, implying that $P = \{1, 2\}$, that $r_1 = r_2 = 1$, that $\ell = 2$, that $Q = \{0, 1\}$, and that $S = \{1, \mu, z\}$ where (this time) $\mu = (01]$ and $1 = (0)(1)$ and $z = (0]$ are members of C_Q.

Then, since $\eta \in C(\eta)$, we obtain (f_η, t_η) by applying (20.1) and (20.3):

$$\begin{aligned} a_{10}\eta = a_{11} \quad &\Rightarrow \quad 1t_\eta = 1 \quad \text{and} \quad 1f_\eta = \mu^1\varepsilon_{r_1} = \mu^1\varepsilon_1 = \mu \circ (0)(1) = \mu \\ a_{20}\eta = a_{21} \quad &\Rightarrow \quad 2t_\eta = 2 \quad \text{and} \quad 2f_\eta = \mu^1\varepsilon_{r_2} = \mu^1\varepsilon_1 = \mu \circ (0)(1) = \mu \end{aligned}$$

Thus, $\eta \mapsto (\mu\mu, (1)(2)) \mapsto [(\mu\mu, (1)(2))]$, and we calculate the product in $\mathcal{W}$ by multiplying representatives and using the congruence $\sim$:

$$\begin{aligned} [(f_\eta, t_\eta)]^2 = [(f_\eta, t_\eta)^2] &= [(f_\eta f_\eta^{t_\eta}, t_\eta^2)] \\ &= [(\mu^2\mu^2, (1)(2))] = [(zz, (1)(2))] = [(zz, (1](2])]. \end{aligned}$$

§22 Orders of Centralizers in C_n

When $\alpha \in C_n$ is a chart that is also a permutation, then (from the previous chapter) the order $|C(\alpha)|$ of the centralizer $C(\alpha)$ is a function of the lengths of the circuits that appear in the path decomposition of α. In this section, we show that when $\eta \in C_n$ is nilpotent, then $|C(\eta)|$ depends only on the lengths of the proper paths appearing in the path decomposition of η.

We begin by assuming $\eta \in C_n$ has the path decomposition (18.1). For each i, $1 \leq i \leq m$, let $s_i = r_i + 1$ be the length of $(a_{i0} \cdots a_{ir_i}]$ and define $m_{ij} = \min\{s_i, s_j\}$. Moreover, for each $\sigma \in C_m$, let $\Pi_{i \in \mathbf{d}\sigma} m_{i,i\sigma} = \Pi_i m_{i,i\sigma}$ denote the product whose factors are $m_{i,i\sigma}$ for $i \in \mathbf{d}\sigma$. When $\sigma = (1]$, i.e., when σ is the empty chart, define $\Pi_i m_{i,i\sigma} = 1$.

Then from Theorem 10.1, the order of the centralizer $C(\eta)$ is

$$|C(x)| = \sum_{\sigma \in C_m} (\Pi_i m_{i,i\sigma})$$

where, for the given σ-matching "$i \leftrightarrow i\sigma$" of the proper paths $(a_{i0} \cdots a_{ir_i}] \leftrightarrow (a_{i\sigma,0} \cdots a_{i\sigma,r_{i\sigma}}]$; the number of ways to map an initial segment of $(a_{i0} \cdots a_{ir_i}]$ onto a final segment of $(a_{i\sigma,0} \cdots a_{i\sigma,r_{i\sigma}}]$ is the minimum $m_{i,i\sigma}$ of the lengths s_i and $s_{i\sigma}$.

For example, consider the $\eta = (12](34]$ featured in §17, where it was claimed that $|C((12](34])| = 17$. In that case, $s_1 = 2$, $s_2 = 2$, $m_{11} = 2$, $m_{22} = 2$, $m_{12} = 2$, $m_{21} = 2$, and $C_2 = \{(1], (1)(2], (1](2), (12], (21], (1)(2), (12)\}$. We then calculate

$$|C\big((12](34]\big)| = 1 + m_{11} + m_{22} + m_{12} + m_{21} + m_{11}m_{22} + m_{12}m_{21} = 17.$$

And when $\eta = (12](345]$; we have

$$s_1 = 2,\ s_2 = 3,\ m_{11} = 2,\ m_{22} = 3,\ m_{12} = m_{21} = 2,$$

and we calculate that

$$|C\big((12](345]\big)| = 1 + m_{11} + m_{22} + m_{12} + m_{21} + m_{11}m_{22} + m_{12}m_{21} = 20.$$

§23 Comments

The constructions and results in this chapter are due to the author [4]. As in the case of the centralizer of a permutation, a few comments on the wreath product are in order.

It should be observed that the wreath product S wr C_m (§18) is of the same ilk as the wreath product Z_ℓ^z wr C_m (§14) used in Chapter 4. That is, they both have their first component semigroup as a semigroup with a zero z, and they share the same equation that defines their multiplication:

$$(f,t)(g,u) = (fg^t, tu) \quad \text{where} \quad p(fg^t) = \begin{cases} (pf)(pt)g & \text{if } p \in \mathbf{d}t \\ z & \text{if } p \notin \mathbf{d}t. \end{cases}$$

This kind of product is essentially the one found in Chapter 1 of Eilenberg [1] or Chapter 4 of Lallement [1].

In comparing the permutation case (Chapter 4) with the nilpotent case (Chapter 5), there is the natural question of whether $\mathcal{W}$ could be replaced with a wreath product, so both cases could be viewed as wreath products. This open question was first posed by Lallement [2] (I have adjusted his notation to conform to the notation used here and any misinterpretation is my fault.):

> In view of the nature of the congruence $\sim$ is it not possible to obtain $\mathcal{W}$ as a wreath product of a transformation semigroup X and of a quotient of a transformation semigroup Y? ... It would be nicer to have $C(\alpha)$ as a wreath product all the time.

CHAPTER 6

Alternating Semigroups

We extend the idea of an even permutation to that of even chart. The even charts in C_n form the alternating semigroup $A_n^c \subset C_n$. As expected, $A_n^c \cap S_n = A_n$, where A_n denotes the alternating group. Moving down one rank, i.e., for charts in C_n of rank $n-1$, we find that exactly half of these charts are members of A_n^c. And below rank $n-1$, we find that all such charts are members of A_n^c.

In addition to specifying its members, we also specify several generating sets. The approach runs parallel to the A_n case, yielding a companion of the "4-group" ($\subset A_4$), namely, the "61-semigroup" ($\subset A_4^c$). The resulting theory dovetails very satisfactorily with the group case — the 61-semigroup is the collection of restrictions of the permutations in the 4-group, and, when $n \geq 5$, the alternating semigroup A_n^c is the collection of restrictions of members of A_n.

§24 Even Charts

For $\alpha \in C_n$ and $m \in \mathbf{d}\alpha$, recall that "α moves m" when $m\alpha \neq m$, and that "α fixes m" otherwise. For instance, recalling that $N = \{1, 2, \ldots, n\}$, the transposition $(i, j) \in S_n$ moves i and j, while fixing each $k \in N - \{i, j\}$, illustrating the convention that whenever a cycle fixes $k \in N$, then the 1-cycle "(k)" does not explicitly appear in the notation.

Similarly, we opt for an alternative usage of the proper path notation: We shall use "$(i, j]$" to denote a rank $n-1$ chart in C_n that fixes each $k \in N - \{i, j\}$, even though none of the 1-circuits "(k)" explicitly appear. In other words, we have

$$(ij] = (ij](j_1) \cdots (j_{n-2}),$$

where $\{j_1, \ldots, j_{n-2}\} = N - \{i, j\}$. Furthermore, in this case we call $(ij]$ a *semitransposition*, the name indicating that $(ij]$ may be obtained from a transposition (i, j) by removing the "arrow" $j \mapsto i$, which is one-half of the two "move-arrows."

With the idea of semitransposition, it is not difficult to get at the facts. For $2 \leq r \leq n$, consider the factorization of $(12 \cdots r) \in S_n$ into transpositions

$$(123 \cdots r) = (r-1, r) \circ (r-2, r-1) \circ \cdots \circ (1, 2),$$

and the analogous factorization, replacing each ")" with "]",

$$(12\cdots r] = (r-1,r]\circ(r-2,r-1]\circ\cdots\circ(1,2] \tag{24.1}$$

of $(12\cdots r] = (12\cdots r](r+1)\cdots(n) \in C_n$ into semitranspositions. In contrast to (24.1), however, certain factorizations of r-cycles may not induce a corresponding factorization for charts. For example,

$$(123\cdots r) = (r,1)\circ(r,2)\circ\cdots\circ(r,r-1),$$

but

$$(12\cdots r] = (12\cdots r](r+1)\cdots(n) \neq (r,1]\circ(r,2]\circ\cdots\circ(r,r-1]. \tag{24.2}$$

The trouble is that the composition on the right side of (24.2) is a rank $(n-r+1)$ chart moving r to 1 and fixing all $k>r$. It is not this failure of (24.2), but rather the success of (24.1) that is important. Before showing why, we need several definitions.

A *transpositional* is a chart that is either a transposition (i,j) or a semitransposition $(i,j]$. A chart is *even* if it is a product of an even number of transpositionals. For $n=1$, the alternating semigroup A_1^c is C_1. For $n>1$, the *alternating semigroup* $A_n^c \subset C_n$ is the subsemigroup of even charts. It is clear that A_n^c is indeed a subsemigroup of C_n, and, since the inverse of a transpositional is a transpositional, A_n^c is also an inverse subsemigroup.

§25 Members of A_n^c

Clearly, A_2^c contains all idempotents in C_2. From Theorem 25.1 (below), every member of A_2^c is an idempotent. Thus, we shall generally be interested in A_n^c where $n \geq 3$.

Unlike permutations, some charts are both "even" and "odd," e.g.,

$$(12](3) = (12]\circ(13] = (12]\circ(23]\circ(32].$$

When n is odd, however, the rank $n-1$ chart $(12\cdots n]$ is even and not odd, which is a fundamental result that also follows from 25.1 (below).

To understand 25.1, some new terminology and additional notation are required. First, we let J_k, $0 \leq k \leq n$, denote the charts in C_n of rank k. Then clearly,

$$C_n = J_0 \cup J_1 \cup J_2 \cup \cdots \cup J_{n-2} \cup J_{n-1} \cup J_n.$$

Furthermore, we let $\mathcal{N}_1$ denote the nilpotent elements in C_n of rank $n-1$. In terms of path notation, the elements in $\mathcal{N}_1$ are precisely the conjugates of

$(12\cdots n]$. To relate these nilpotents to permutations, we define the *completion $\bar{\alpha}$ of a chart $\alpha \in C_n$*: In the path decomposition of α, we (mechanically) replace each instance of "]" with ")". So the completion operation is a map from C_n onto S_n that extends the identity $S_n \to S_n$.

25.1 Theorem (**Even charts of rank** $n-1$)

Let $\alpha \in C_n$ have rank $n-1$, i.e., $\alpha \in J_{n-1}$. Then $\alpha \in A_n^c$ if and only if its completion $\bar{\alpha}$ is an even permutation of N.

Proof. ($\Rightarrow$) Let $\alpha \in A_n^c$. Then α has a factorization $\alpha = \alpha_1 \circ \cdots \circ \alpha_k$ into an even number k of transpositionals. Since the rank of α is $n-1$, it is exactly "one arrow short" of a permutation. Likewise, each transpositional α_i is at most one arrow short of a permutation. To see the connection, suppose $(x_1, x_k) \notin \alpha$ is the missing arrow. Then since $\alpha = \alpha_1 \circ \cdots \circ \alpha_k$, there is a unique list of "arrows" (x_1, x_2), (x_2, x_3), $\ldots$, (x_{k-1}, x_k) containing all missing arrows of those α_i that are semitranspositions. It follows that the completion $\bar{\alpha}$ equals the composition of the completions $\bar{\alpha}_1 \circ \cdots \circ \bar{\alpha}_k$. This shows that $\bar{\alpha}$ is even. ($\Leftarrow$) Conversely, let $\bar{\alpha}$ be even. If $\bar{\alpha}$ is an n-cycle, then n must be odd, and α must be a proper path of length n. A factorization as in equation (24.1) above then shows that $\alpha \in A_n^c$. Next suppose $\bar{\alpha}$ is not an n-cycle. Then $\bar{\alpha}$ must, in its cycle decomposition, have an even number of cycles of even length. With this cycle structure of $\bar{\alpha}$, the path decomposition of α must also have an even number of paths of even length. Factoring each of the (even number of) even-length paths into an odd number of transpositionals, and the remaining odd-length paths into an even number of transpositionals, the desired result follows. □

25.2 Theorem (**All charts of rank** $< n-1$ **are even.**)

If $\alpha \in C_n$ has rank at most $n-2$, then $\alpha \in A_n^c$.

Proof. If the rank of α is $\leq n-2$, then the path decomposition of α must have the following form

$$\alpha = (a_{11} \cdots a_{1r_1}](a_{21} \cdots a_{2r_2}](\cdots] \cdots (\cdots](\cdots) \cdots (\cdots),$$

i.e., at least two "]'s" must appear. Then, since α can be factored into transpositionals, and since

$$\alpha = \alpha \circ (a_{11}, a_{21}],$$

we see that α is both even and odd, *a fortiori*, even. □

25.3 Proposition (**There are enough even charts.**)

Let $n \geq 3$ and let $A, B \subset \{1, 2, \ldots, n\}$ have the same size, i.e., $|A| = |B|$. Then there is a chart $\alpha \in A_n^c$ whose domain is A and whose range is B.

Proof. The only troublesome case occurs when $|A| = |B| = n - 1$. So let $A = \{x, i_1, \ldots, i_{n-2}\}$ and $B = \{y, i_1, \ldots, i_{n-2}\}$. If $x = y$, i.e., $A = B$, then let $\alpha = \varepsilon$ be the idempotent whose domain is A. If $x \neq y$, then $A \neq B$, and when n is odd, choose α to be proper (odd-length) path $(x, i_1, \ldots, i_{n-2}, y]$. Otherwise, n is even and we may choose α to be the join $\varepsilon\eta$ of the rank-one idempotent $\varepsilon = (i_1)$ with the proper (odd-length) path $\eta = (x, i_2, \ldots, i_{n-2}, y]$. $\square$

It follows from 25.1 and 25.2 that if J_{n-1}^e denotes the set of rank $n-1$ charts whose completions are even, then

$$A_n^c = A_n \cup J_{n-1}^e \cup (J_{n-2} \cup \cdots \cup J_0).$$

And that this partition (according to rank) is the Green's $\mathcal{D}$-class partition of A_n^c follows from 25.3. We can say more. Since a rank $n-1$ group $\mathcal{H}$-class of A_n^c clearly contains exactly half of the members in the corresponding $\mathcal{H}$-class of C_n, we see that J_{n-1}^e contains exactly half of the members of J_{n-1}.

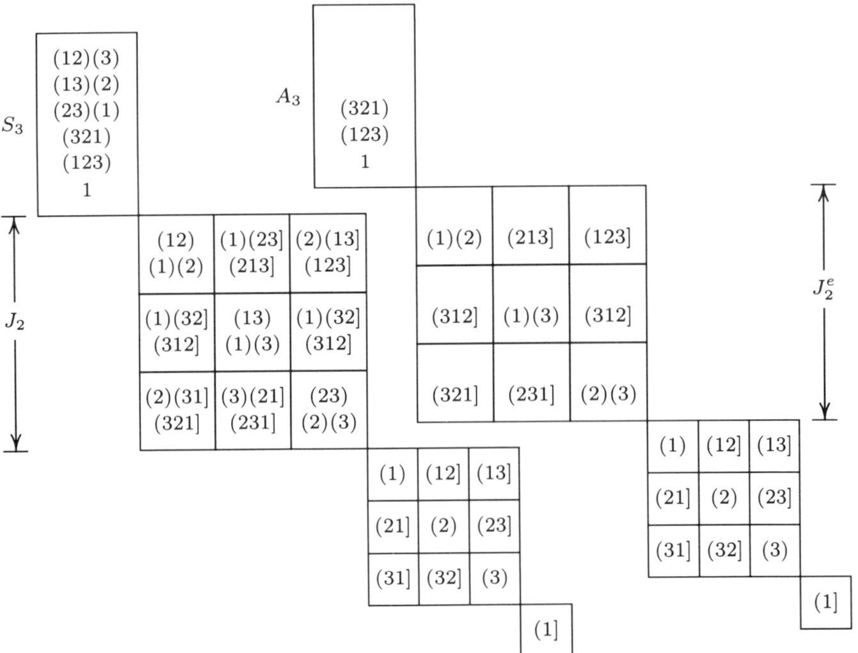

Figure 25.4. Comparing C_3 with the alternating semigroup A_3^c.

As a consequence, the order of A_n^c is

$$|A_n^c| = \frac{1}{2}n! + \frac{1}{2}n^2(n-1)! + \sum_{k=0}^{n-2} \binom{n}{k}^2 k!.$$

For example, the 22 members of A_3^c are pictured in Figure 25.4.

§26 Generators of Alternating Semigroups

To make the comparison of A_n and A_n^c most obvious, we first extend the idea of *type*. Let $\alpha \in C_n$ have the following path decomposition

$$\alpha = (a_{11}a_{12}\cdots a_{1u_1})\cdots(a_{m1}\cdots a_{mu_m})(b_{11}b_{12}\cdots b_{1v_1}]\cdots(b_{k1}\cdots b_{kv_k}].$$

Then the arrangement of the lengths of the circuits in increasing order of magnitude $u_1 \leq u_2 \leq \cdots \leq u_m$, and the arrangement of the lengths of the proper paths in increasing order of magnitude $v_1 \leq v_2 \leq \cdots \leq v_k$, is the *type of the chart* α. For example, the path (123)(45)(5678](9] in C_9 has type (2,3;1,4). Also, since the permutation $\alpha = (12)(34) \in C_5$ has type (1,2,2), we shall agree to suppress the "1" whenever it indicates the presence of a 1-circuit, making (2,2) the type of α. Extending this convention to charts, e.g., either (1;1,2,2) or (;1,2,2) will denote the type of (12](34](6] $\in C_5$.

Three basic facts concerning generators of A_n are standard: (i) *For* $n \geq 3$, *the alternating group* A_n *is generated by the totality of* 3-*cycles*; (ii) *If* $n \geq 5$, A_n *is generated by the totality of permutations of type* $(2,2)$; (iii) *If* $n = 4$, *the set of permutations of type* $(2,2)$ *generates a normal subgroup, namely the* 4-*group, of order* 4 *in* S_4.

26.1 Theorem (Generators of A_n^c)

(i) *If* $n \geq 3$, *the alternating semigroup* A_n^c *is generated by the totality of* 3-*paths*. (ii) *If* $n \geq 5$, A_n^c *is generated by the totality of charts of types* $(2,2)$, $(2;2)$. (iii) *If* $n = 4$, *the set* T, *which contains both type* $(2,2)$ *and type* $(2;2)$ *charts, generates a semigroup* $\langle T\rangle$ *of order* 61. *Furthermore,* $\langle T\rangle$ *is the collection of restrictions of the* 4-*group permutations. And, for* $n \geq 5$, A_n^c *is the collection of restrictions of the even permutations.*

Proof. By definition, the semigroup A_n^c is generated by elements that are products of two transpositionals. These two-transpositional-products, or their inverses, are organized into three columns; each column having a fixed number (4 or 3 or 2) of differing letters representing distinct numerals:

$$\begin{array}{lll}
(ij)\circ(km) & (ij)\circ(ik) & (ij)\circ(ij) \\
(ij]\circ(km) & (ij]\circ(ik) & (ij]\circ(ij) \\
 & (ji]\circ(ik) & \\
(ij]\circ(km] & (ij]\circ(ik] & (ij]\circ(ij] \\
 & (ji]\circ(ik] & (ji]\circ(ij] \\
 & (ij]\circ(ki]. &
\end{array}$$

If different letters represent distinct numerals, then for the first column,

$$\begin{aligned}
(ijk)\circ(imk) &= (ij)\circ(km) \\
(ijk)\circ(imk] &= (ij]\circ(km) \\
(ijm]\circ(ikm] &= (ij]\circ(km].
\end{aligned}$$

For the second and third columns we have

$$\begin{array}{rr}
(kij) = (ij)\circ(ik) & (ijk)^3 = (ij)\circ(ij) \\
(kij] = (ij]\circ(ik) & (ikj]\circ(ikj]^{-1} = (ij]\circ(ij) \\
(jki] = (ji]\circ(ik) & \\
(ikj]^2 = (ij]\circ(ik] & (ikj]\circ(ijk] = (ij]\circ(ij] \\
(jik]^2 = (ji]\circ(ik] & (jki]\circ(jki]^{-1} = (ji]\circ(ij] \\
(kij] = (ij]\circ(ki]. &
\end{array}$$

This proves (i). If $n \geq 5$, then for any given (ijk) and $(ijk]$, we can find ℓ and m such that

$$\begin{aligned}
(ijk) &= (ij)(\ell m)\circ(\ell m)(ik) \\
(ijk] &= (ij)(\ell m)\circ(\ell m)(ik].
\end{aligned}$$

This proves (ii). If $n = 4$, then the members of T are just the 15 charts of types (2,2),(2;2):

$$\begin{array}{lll}
(12)(34) & (12)(34] & (12](34) \\
 & (12)(43] & (21](34) \\
(13)(24) & (13)(24] & (13](24) \\
 & (12)(42] & (21](24) \\
(14)(23) & (14)(23] & (14](23) \\
 & (14)(32] & (41](23).
\end{array}$$

Table 26.2. The 61-semigroup.

Path Form	**Number**	**Rank**
$(ij)(k\ell)$	$\frac{1}{2} \times \binom{4}{2} = 3$	4
$(i)(j)(k)(\ell)$	$\binom{4}{4} = 1$	4
$(ij)(k\ell]$	$\binom{4}{2} \times 2 = 12$	3
$(i)(j)(k)(\ell]$	$\binom{4}{3} = 4$	3
$(ij)(k](\ell]$	$\binom{4}{2} = 6$	2
$(ij](k\ell]$	$\frac{1}{2}\binom{4}{2} \times 4 = 12$	2
$(i)(j)(k](\ell]$	$\binom{4}{2} = 6$	2
$(ij](k](\ell]$	$\binom{4}{2} \times 2 = 12$	1
$(i)(j](k](\ell]$	$\binom{4}{1} = 4$	1
$(i](j](k](\ell]$	$\binom{4}{0} = 1$	0
	Order $= 61$	

Next, let α_{ij} denote $(ij](k\ell)$; β_{ij} the chart $(ij)(k\ell)$; and $1'_K$ the idempotent in C_4 whose domain is the complement of $K \subset \{1,2,3,4\}$. First, use $\beta_{ij}^2 = 1'_\emptyset = 1$ and $\alpha_{k\ell} \circ \alpha_{\ell k} = (i)(j)(k)(\ell]$ to show every idempotent in C_4 is also in $\langle T \rangle$. Second, use $\alpha_{ij} = 1'_j \circ \beta_{ij} = \beta_{ij} \circ 1'_i$ to show that any $\rho \in \langle T \rangle$ has a factorization

$$\rho = 1'_K \circ \beta,$$

where β is a product of (2,2) permutations. Thus, either $\beta = 1$, or $\beta = (ij)(k\ell)$. Since the 4-group $V \subset C_n$ is generated by the (2,2) permutations, the 61-semigroup $\langle T \rangle$ consists of the restrictions of 4-group permutations (see Table 26.2 and Figure 39.5). This proves (iii) and the stated characterization of $\langle T \rangle$.

Finally, let $n \geq 5$. Then by (ii), A_n^c is generated by the (2,2),(2;2) charts. Hence, as with $\langle T \rangle$, any $\rho \in A_n^c$ has a factorization $\rho = 1'_K \circ \beta$, where $K \subset \{1,2,\dots,n\}$ and $\beta \in A_n$. Therefore, any $\rho \in A_n^c$ is a restriction of some $\beta \in A_n$. Conversely, since any idempotent $1'_K$ is even and any $\beta \in A_n$ is even, their product ρ must also be an even chart. Hence, any restriction of any $\beta \in A_n$ is in A_n^c. This proves the stated characterization of A_n^c. □

The representation $\rho = 1'_K \circ \beta$ (in the proof of 26.1) also sheds light on Theorem 25.1: Indeed, from this representation $\alpha \in A_n^c \cap J_{n-1}^e$ if and only if $\alpha = 1'_i \circ \beta$ for some $i \in \{1,2,\dots,n\}$ and some $\beta \in A_n$. In these terms, Theorem 25.1 states that the completion of each $1'_i \circ \beta$, for $\beta \in A_n$, is an even permutation. For example, suppose $n \geq 5$ is odd, then

$$\rho = 1'_i \circ (1, 2, \cdots, i-1, i, i+1, \cdots, n) = (i+1, i+2, \cdots, n, 1, 2, \cdots, i-1, i],$$

and its completion $\bar{\rho} = (1 \cdots n)$ is indeed an even permutation. More generally, "cut" any permutation $\alpha = (\cdots ik \cdots) \cdots$ by premultiplying by $1'_i$ to obtain $\alpha' = (k \cdots i] \cdots$. Then "paste" (or complete) the resultant α' between i and k, i.e., change the "i]" to "i)"and we will be back to α. The key is to make only *one* cut. Thus, 25.1 is a statement of this surgical invariance where exactly one "cut" is allowed. After one of a bunch of rubber bands is cut, the one that was cut is known. After two cuts are made however, it is not known whether one band was cut into two pieces (in which case the original cyclic order of the band cannot be determined because of the induced two-fold symmetry), or whether two bands were cut. In this sense, Theorem 25.1 is the "one-cut Theorem" (rank $n-1$), while Theorem 25.2 is the "several-cut Theorem" (rank $< n-1$).

§27 Comments

Along with cycle notation and even permutations, Cauchy [1] and [2] introduced the alternating groups A_n in 1815. Nearly two centuries later (1987), R. P. Sullivan [1, page 330] was evidently the first to advance a concept of an *even transformation*:

> "If $\beta \in \mathcal{I}_X$ (C_X), we call β an *even transformation* if it is an even permutation of its domain, or a chain with even rank, or a disjoint union of an odd permutation and a chain with odd rank, or a disjoint union of any of the previous three types of transformations. It is important to note that for the purpose of this definition non-empty idempotents in $\mathcal{I}_X$ are to be regarded as even permutations ... ."

Independent of Sullivan, in 1992 the author [5] introduced both even charts and alternating semigroups as they are defined in §24. That these alternating semigroups had connections with other semigroups was immediately recognized. For example, A_n^c is connected to the semigroup $\langle \mathcal{N}_1 \rangle$ generated by the set $\mathcal{N}_1 \subset C_n$ of nilpotents of rank $n-1$. To be precise, for $n \geq 3$, let I_{n-1} denote the ideal in C_n consisting of all charts of rank $\leq n-1$, then

$$\langle \mathcal{N}_1 \rangle = \begin{cases} A_n^c - A_n & \text{if } n \text{ is odd;} \\ I_{n-1} & \text{if } n \text{ is even.} \end{cases}$$

Gomes and Howie [2] had introduced and characterized $\langle \mathcal{N}_1 \rangle$ in 1987. In fact, other results in Gomes and Howie [2] relate to A_n^c: They proved that $(1 \cdots r](r+1) \cdots (n)$ is even when n is odd (their Lemma 3.10), they introduced the notion of *completion*, which, for the rank $n-1$ charts in C_n, agrees with the one given in §24, and they defined the idea of *tidy product*, which was used by the author [5] in an alternative proof of 25.1.

For a development of the group theory (with semigroup theory removed) that runs closely parallel to §26, see Suzuki [1, p 293]. Following the study of congruences of C_n by Scheiblich [1], Lipscomb [3] showed that the congruences of A_n^c also form a chain. The author originally calculated that the 61-semigroup had less than 61 members! The remaining members were discovered by M. Hermann's computer program.

The label "alternating semigroup" has also been used by J. Dénes [1] to name a subsemigroup K_n of the full transformation semigroup T_n. To construct K_n, remove S_n from T_n, and then add back A_n, i.e., the semigroup $K_n = (T_n - S_n) \cup A_n$.

CHAPTER 7

S_n-normal Semigroups

With transpositionals, the "evenness argument" showing $\alpha^{-1}A_n\alpha \subset A_n$ for each $\alpha \in S_n$ extends to $\alpha^{-1}A_n^c\alpha \subset A_n^c$ for each $\alpha \in C_n$. In other words, A_n^c is (chart) self-conjugate. Moreover, since A_n^c is full (contains all idempotents of C_n), it is both full and self-conjugate, making it normal. Normal semigroups are important because they are fundamental to determining congruences on C_n (Chapter 8).

It turns out that a full subsemigroup T is normal in C_n whenever it is merely permutation self-conjugate, i.e., $\alpha^{-1}T\alpha \subset T$ for each permutation $\alpha \in S_n$. The permutation self-conjugate (not necessarily full) subsemigroups of C_n are the S_n-normal semigroups. In this chapter, we classify the S_n-normal semigroups.

Following the design and spade work in §§28–32, we present the classification in §33 ($n \geq 5$) and §34 ($n < 5$). Roughly speaking, S_n-normal semigroups are expressed in terms of the following: normal subgroups of symmetric groups, ideals I_r comprised of charts of rank $\leq r$, semilattices E_m of idempotents of rank $\leq m$, alternating semigroups A_n^c, and the X_n-semigroups (§29), which only appear when n is even.

§28 Example, Lemma, Overview, Notation

For $n = 2$, we have 10 S_2-normal semigroups (Figure 28.1). To understand these semigroups, let "1" denote the identity (1)(2) of S_2. Then the S_2-normal semigroups derive from:

(1) normal subgroups of S_2, which are $\{1\}$ and S_2;
(2) ideals of C_2, which are $I_0 = \{(1]\}$, $I_1 = I_0 \cup \{(1), (2), (12], (21]\}$, and $I_2 = C_2$;
(3) semilattices of idempotents, which are $E_1 = \{(1), (2), (1]\}$ and $E_2 = E_1 \cup \{(1)(2)\}$;
(4) the alternating semigroup $A_2^c = E_2$; and
(5) the X_2-semigroup $X_2 = I_1$.

Only for $n = 2$ is the alternating semigroup A_2^c a semilattice and the X_2-semigroup an ideal.

To motivate our approach in the general case, let T be S_2-normal and

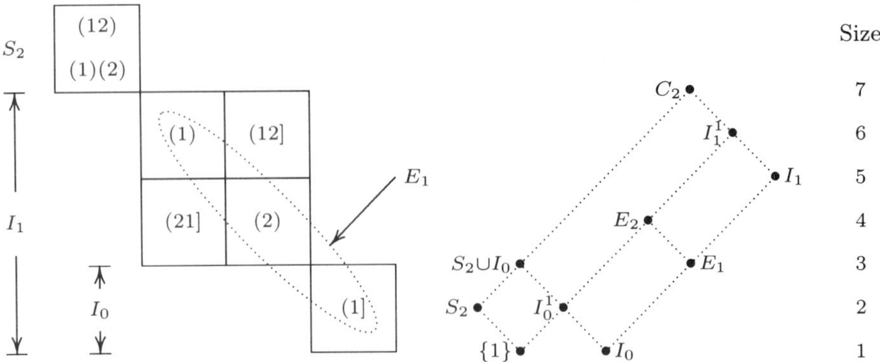

Figure 28.1. Egg-box picture of C_2 and lattice of S_2-normal semigroups.

suppose the nonidempotent $(12] \in T$. Then $I_1 \subset T$:

$$(12] \in T \;\Rightarrow\; (21] \in T \;\Rightarrow\; (1),(2) \in T \;\Rightarrow\; (1] \in T.$$

The first implication follows since (21] is conjugate to (12], the second and third by considering products of conjugate charts, e.g., $(12] \circ (21] = (1)$ and $(1)(2] \circ (2)(1] = (1]$.

This exercise for $n = 2$ illustrates equation (30.1), which is fundamental for larger values of n. The companion equation (30.2) is likewise fundamental. In addition to these equations, we shall frequently multiply conjugate charts by referring to the following lemma. (Its proof amounts to performing the indicated multiplications and is therefore omitted.)

28.2 Lemma

Let $\ell \geq 3$. Then

(1) $(1,\cdots,\ell] \circ (\ell-1,\ell,\ell-2,\cdots,2,1] = (1)\cdots(\ell-3)(\ell-1,\ell-2,\ell]$,

(2) $(1,\cdots,\ell) \circ (\ell-1,\ell,\ell-2,\cdots,2,1) = (1)\cdots(\ell-3)(\ell-1,\ell-2,\ell)$,

(3) $(xy](12)(3)\lambda \circ (3x](12)(y)\lambda^{-1} = (3xy](1)(2)(\lambda \circ \lambda^{-1})$,

(4) $(xy](12)(3]\lambda \circ (y3](12)(x)\lambda^{-1} = (x3](y](1)(2)(\lambda \circ \lambda^{-1})$,

(5) $(xy](12](3)\lambda \circ (yx](31](2)\lambda^{-1} = (312](y](x)(\lambda \circ \lambda^{-1})$,

(6) $(xy](12](3]\lambda \circ (y3](2x](1]\lambda^{-1} = (1x3](y](2](\lambda \circ \lambda^{-1})$,

(7) $(xy](12)(34)\sigma \circ (1x](y2)(34)\sigma^{-1} = (1y](2x)(3)(4)(\sigma \circ \sigma^{-1})$,

(8) $(xy](12](34)\sigma \circ (yx](31](24)\sigma^{-1} = (32](14)(y](x)(\sigma \circ \sigma^{-1})$,

(9) $(xy](1)\delta \circ (y1](x)\delta^{-1} = (x1](y](\delta \circ \delta^{-1})$,

(10) $(xy](12]\rho \circ (x1](y2]\rho^{-1} = (x2](y](1](\rho \circ \rho^{-1})$, *and*

(11) $(x](12)\tau \circ (2](1x)\tau^{-1} = (2x](1](\tau \circ \tau^{-1})$.

With this technical lemma and equations (30.1) and (30.2), we have the tools for building the classification. So let us now overview the plan by considering S_n-normal semigroups that are not groups. These partition into two kinds — those with and those without nonidempotents of rank $< n$. The latter kind are easy to analyze. As for the former kind, when T has nonidempotents of rank $< n$, we focus on the maximum rank $r < n$ for which a nonidempotent $\alpha \in T$ exists. At ranks less than r, we show (§31) that the ideal I_{r-1} of all charts of rank $< r$ is included in T. At rank r, we show (§32) that T must be one of the following:

(1) J_r (the $\mathcal{J}$-class of rank r),
(2) $A_n^c \cap J_r$ (only when $r = n-1$),
(3) $X_n \cap J_r$ (only when n is even and $r = n/2$), or
(4) a union of normal subgroups of maximal groups in J_r.

To begin the spade work, recall (Chapter 6) that we used "(12]" to denote the transpositional $(12](3)\cdots(n) \in C_n$, whose rank is $n-1$. We now adopt the practice of indicating rank with subscripts. For example, whenever it is necessary to distinguish between the proper 2-path (12] and the rank $n-1$ transpositional (12], we shall denote the latter as "$(12]_{n-1}$".

More generally, for $\ell \leq r+1 \leq n$, a *proper ℓ-path of rank* r has domain of size r and path decomposition

$$(i_1, i_2, \ldots, i_\ell]_r = (i_1, i_2, \ldots, i_\ell]p_1p_2\cdots p_{n-\ell},$$

where each p_t is either $(j_t]$ or (j_t). Similarly, for $\ell \leq r \leq n$, any chart whose domain has size r and whose path decomposition has the form

$$(i_1, i_2, \ldots, i_\ell)_r = (i_1, i_2, \ldots, i_\ell)p_1p_2\cdots p_{n-\ell},$$

where each p_t is either $(j_t]$ or (j_t) is an *ℓ-circuit of rank* r. In particular, a 1-circuit of rank r is an idempotent, and the idempotents of rank r generate the *semilattice E_r of idempotents of rank* $\leq r$. More precisely, if for each $K = \{i_1 \ldots, i_r\} \subset N$ of size $r \geq 0$, we use "1_K" to denote the idempotent whose join of 1-paths is $1_K = (i_1)\cdots(i_r)(j_1]\cdots(j_{n-r}]$, then for each $r < n$, we define

$$E_r = \langle 1_K \mid K \subset N; |K| = r\rangle. \tag{28.3}$$

In terms of the notation $|\alpha|$ (for the rank of $\alpha \in C_n$), we have $E_r = \{\, \varepsilon \in C_n \mid \varepsilon^2 = \varepsilon \text{ and } |\varepsilon| \leq r \,\}$. We shall also use idempotents to restrict charts, e.g., the *restriction* $\alpha|_K$ of α to $K \subset \mathbf{d}\alpha$ is

$$\alpha|_K = 1_K \circ \alpha = \alpha \circ 1_M = 1_K \circ \alpha \circ 1_M \tag{28.4}$$

where $M = \{\, k\alpha \,|\, k \in K \,\}$.

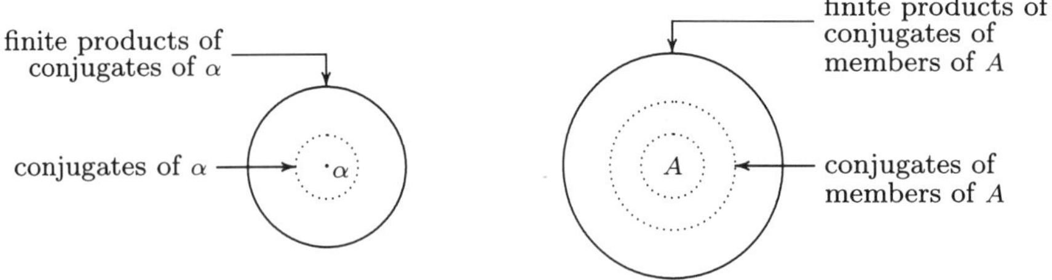

Figure 28.5. Constructing S_n-normalizers $\langle\, \alpha\, ;\, S_n\, \rangle$ and $\langle\, A\, ;\, S_n\, \rangle$.

And finally, we define an S_n-normalizer: For $A \subset C_n$, the S_n*-normalizer* $\langle A; S_n\rangle$ of A is the smallest S_n-normal semigroup having A as a subset.

As indicated in Figure 28.5, we have $\langle A; S_n\rangle = \langle \{\, \beta^{-1}\alpha\beta \mid \beta \in S_n;\, \alpha \in A\,\}\rangle$, implying that when A is (permutation) self-conjugate ($\beta^{-1}A\beta \subset A$ for all $\beta \in S_n$), then $\langle A; S_n\rangle = \langle A\rangle$. In particular, if $r < n$ and $A = \{\, \alpha \in C_n \mid |\alpha| = r\}$, then A is (permutation) self-conjugate and $\langle A; S_n\rangle$ is the ideal I_r of all charts in C_n of rank $\leq r$.

§29 X Semigroups

Consider the subsemigroup $X_4 = \langle \beta^{-1}(12](34]\beta \mid \beta \in S_4\,\rangle$ of C_4 pictured in Figure 39.5. To characterize the generators of X_4, recall that conjugacy of charts equates to same path structure, i.e.,

$$\alpha = \beta^{-1}(12](34]\beta \Leftrightarrow \alpha = (xy](uv] \Leftrightarrow |\mathbf{d}\alpha| = 2 \text{ and } \mathbf{d}\alpha \cap \mathbf{r}\alpha = \emptyset.$$

As a consequence, if H_α denotes the $\mathcal{H}$-class containing a generator $\alpha = (xy](uv]$, then, since

$$\begin{aligned} \beta \in H_\alpha &\Leftrightarrow \mathbf{d}\beta = \mathbf{d}\alpha \text{ and } \mathbf{r}\beta = \mathbf{r}\alpha \\ &\Rightarrow |\mathbf{d}\beta| = 2 \text{ and } \mathbf{d}\beta \cap \mathbf{r}\beta = \emptyset \\ &\Rightarrow \beta = \gamma^{-1}(12](34]\gamma \in X_4, \end{aligned}$$

$H_\alpha \subset X_4$ is an $\mathcal{H}$-class of generators. In particular, forming the J_2 egg-box via dictionary ordering (12,13,14,23,24,34) of $\mathcal{H}$-classes, we see that the generator $\mathcal{H}$-classes are those on the off-diagonal. As for a product of two such $\mathcal{H}$-classes,

$$H_\alpha H_\beta = \begin{cases} H_{\alpha\alpha^{-1}}, & \text{if } \beta \in H_{\alpha^{-1}}\,; \\ \text{a subset of } I_1, & \text{otherwise,} \end{cases}$$

showing that each $\mathcal{H}$-class on the main-diagonal is also a subset of X_4. Moreover, it is easy to show that these main- and off-diagonal $\mathcal{H}$-classes are the only rank 2 $\mathcal{H}$-classes in X_4, i.e., $X_4 \cap J_2$ resembles an "X." If H_{ij}^{kt} denotes the $\mathcal{H}$-class corresponding to domain $\{i, j\}$ and range $\{k, t\}$, then

$$\begin{array}{cccccc} H_{12}^{12} & & & & & H_{12}^{34} \\ & H_{13}^{13} & & & H_{13}^{24} & \\ & & H_{14}^{14} & H_{14}^{23} & & \\ & & H_{23}^{14} & H_{23}^{23} & & \\ & H_{24}^{13} & & & H_{24}^{24} & \\ H_{34}^{12} & & & & & H_{34}^{34} \end{array}$$

illustrates $X_4 \cap J_2$. As for the rest of X_4, since

$$(xy](zt] \circ (x)(y)(t](z] = (xy](t](z]$$

the ideal $I_1 = \{ \alpha \in C_4 \mid |\alpha| \leq 1 \}$ is also a subset of X_4.

In general, i.e., for $n = 2k \geq 2$, we define *the X semigroup on n symbols* as the subsemigroup X_n of C_n generated by charts conjugate to $(12] \cdots (n-1, n]$. To show that $X_n \cap J_k$ resembles an "X," we can use the obvious generalization of our $n = 4$ argument above; and to show that $I_{k-1} \subset X_n$, we may apply 31.2 below.

Finally, since $A = \{ \beta^{-1}(12] \cdots (n-1, n]\beta \mid \beta \in S_n \}$ is a self-conjugate set, $X_n = \langle A; S_n \rangle$ is S_n-normal.

§30 Basic Observations

The classification rests on two observations, namely, that proper 2-paths $(ij]_r$ (of rank $r < n$) generate the ideal $I_r = \{ \alpha \in C_n \mid |\alpha| \leq r \}$, i.e.,

$$I_r = \langle\, \beta^{-1}(12]_r\beta \mid \beta \in S_n \,\rangle = \langle (12]_r; S_n \rangle, \tag{30.1}$$

and that proper 3-paths $(ijk]$ (of rank $n-1$) generate $A_n^c - A_n$, i.e.,

$$A_n^c - A_n = \langle\, \beta^{-1}(123]\beta \mid \beta \in S_n \,\rangle = \langle (123]; S_n \rangle. \tag{30.2}$$

To prove (30.1), first note that

$$\begin{aligned}(1) \cdots (r)(r+1] \cdots (n] = (1, r+1](2) \cdots (r)(r+2] \cdots (n]\circ \\ (r+1, 1](2) \cdots (r)(r+2] \cdots (n],\end{aligned}$$

showing that each rank r idempotent ε is a product of proper 2-paths of rank r. Second, observe that each chart α is a restriction $\beta|_{\mathbf{d}\alpha}$ of some permutation β: If we write $\beta = \tau_1 \circ \tau_2 \circ \cdots \circ \tau_k$ as a product of transpositions τ_i and if the rank of α is $r < n$, then

$$\alpha = \tau_1 \circ \tau_2 \circ \ldots \tau_{k-1} \circ \varepsilon' \circ \tau_k \circ \varepsilon \qquad (\varepsilon = 1_{\mathbf{r}\alpha} \text{ and } \varepsilon' = 1_{\mathbf{d}(\tau_k \circ \varepsilon)}).$$

Suppose $\tau_k = (12)$. If $\varepsilon = (x]\varepsilon_1$ with $x \in \{1,2\}$, then $\tau_k \circ \varepsilon$ is either ε or a proper 2-path of rank r. Otherwise, $\varepsilon = (1)(2)(y]\varepsilon_1$ and $\tau_k \circ \varepsilon$ is a rank r 2-circuit $(12)_r = (12)(y]\varepsilon_1$. But in this case,

$$(12)(y]\varepsilon_1 = (2y](1)\varepsilon_1 \circ (12](y)\varepsilon_1 \circ (y1](2)\varepsilon_1,$$

showing that we can replace $\tau_k \circ \varepsilon$ with a product of rank r proper 2-paths. It now follows from induction that α itself is such a product. Finally, since any chart of rank $k < r$ is a restriction of a rank r chart, we see that (28.3) and (28.4) imply (30.1).

With (30.1) established, we can prove (30.2): Indeed, (30.1) shows that semitranspositions generate $C_n - S_n$, and, since each chart α in $A_n^c - A_n$ is a product of an even number of semitranspositions, $A_n^c - A_n$ is generated by pairs of semitranspositions. So let us organize the possible pairs (excluding unnecessary inverses) into three columns, each having a fixed number (4 or 3 or 2) of differing letters representing distinct numerals:

$$\begin{array}{lll} (ij] \circ (km] & (ij] \circ (ik] & (ij] \circ (ij] \\ & (ji] \circ (ik] & (ji] \circ (ij] \\ & (ij] \circ (ki] & \end{array}$$

Then for the first column, we have

$$(ij] \circ (km] = (ijm] \circ (ikm].$$

For the second and third columns, we have

$$\begin{array}{ll} (ij] \circ (ik] = (ikj]^2 & (ij] \circ (ij] = (ikj] \circ (ijk] \\ (ji] \circ (ik] = (jik]^2 & (ji] \circ (ij] = (jki] \circ (ikj] \\ (ij] \circ (ki] = (kij] & \end{array}$$

which proves (30.2).

§31 Semilattices and Ideals

The simplest kinds of S_n-normal semigroups are the semilattices E_r, where r is such that $0 \le r \le n$. Indeed, we have the following proposition.

31.1 Proposition

Let $n \geq 1$. If $\varepsilon \in C_n$ is an idempotent of rank $r < n$, then $\langle \varepsilon; S_n \rangle = E_r$.

Proof. If $\varepsilon = (i_1)\cdots(i_r)(j_1]\cdots(j_{n-r}]$, then all idempotents of rank r have the same path structure as ε, i.e., are conjugate to ε. Thus, since $r < n$, (28.3) yields $E_r \subset \langle\varepsilon; S_n\rangle$. For the other inclusion, note that any product of rank r idempotents is an idempotent of rank $k \leq r$. □

Next to the semilattices E_r, the ideals $I_r = \{\,\alpha \in C_n \mid |\alpha| \leq r\,\}$, $0 \leq r \leq n$, are the simplest S_n-normal semigroups.

31.2 Proposition

Let $n \geq 5$ and $0 < r < n$. If $\alpha \in C_n$ is nonidempotent of rank r, then $I_{r-1} \subset \langle\alpha; S_n\rangle$.

Proof. Let $T = \langle\alpha; S_n\rangle$. If $r = 1$, then $\alpha = (i_1 i_2]_1 = (i_1 i_2](i_3]\cdots(i_n]$ and $\alpha^2 = 0$, showing $I_0 \subset T$. If $r > 1$, then by (30.1) it suffices to show that T contains a rank $r-1$ proper 2-path η. To that end, since α is not an idempotent, we first observe that its path decomposition contains an ℓ-path with $\ell \geq 2$. If $\ell \geq 3$, then (1) and (2) of 28.2 show that T contains either $\rho = (123]_r$ or $\gamma = (123)_r$. If $\rho \in T$, we let $\eta = \rho \circ 1_{\mathbf{d}\rho}$. If $\gamma \in T$, then $r < n$ implies $\gamma = (x](123)\varepsilon_1$ for some join ε_1 of 1-paths. In this case, let

$$\eta = (x](123)\varepsilon_1 \circ (x](1](2)(3)\varepsilon_1 \circ (1](32x)\varepsilon_1 \circ (1](3](2)(x)\varepsilon_1 = (1x](3](2)\varepsilon_1.$$

So we may now suppose that the decomposition of α contains no paths of length ≥ 3 and that $\ell = 2$. The possibilities, since $n \geq 5$ and $1 < r < n$, reduce to T containing one of the following: (*i*) $(xy](1)\delta$, (*ii*) $(x](12)\tau$ (*iii*) $(xy](12]\rho$, or (*iv*) $(xy](12)(34)\sigma$. Then (*i*) gives the desired η by (9) of 28.2, (*ii*) by (11) of 28.2, (*iii*) by (10) of 28.2, and (*iv*) by (7) of 28.2 and then (9) of 28.2. □

§32 Nonidempotents in S_n-normal Semigroups

In §31 we showed that if α is nonidempotent with $0 < |\alpha| = r < n$, then $I_{r-1} \subset \langle\alpha; S_n\rangle$. At rank r, the calculation of $\langle\alpha; S_n\rangle \cap J_r$ bifurcates into separate cases, namely, the case where the index of α is greater than 1 and the case where the index of α is 1. Moreover, the "> 1 case" further bifurcates into $r = n-1$ and $r < n-1$, with special attention needed when α is conjugate to $(12](34]\cdots(n-1, n]$.

(This flow and branching is organized and illustrated in Figure 32.2. Beginning at the top of Figure 32.2, we shall first follow the right-side branch (index $\alpha > 1$), and then second, returning to the top, we shall follow the left-side branch (index $\alpha = 1$).)

Nonidempotents of Index > 1.

The results in this subsection follow mainly from Lemma 28.2.

32.1 Lemma

Let $n \geq 1$, and let η be the nilpotent part of $\alpha \in C_n$. Then η is also the nilpotent part of some $\beta \in \langle \alpha; S_n \rangle$, where each circuit in the decomposition of β has length at most 2.

Proof. It suffices to show that if $\alpha = \eta(1, 2, \ldots, \ell)\gamma_1$ with $\ell \geq 3$, then there exists a join γ_2 of circuits having length strictly less than ℓ such that $\eta\gamma_2\gamma_1 \in \langle a; S_n \rangle$. So suppose $\ell \geq 3$. Then by (2) of 28.2,

$$\begin{aligned}\eta(1, \cdots, \ell)\gamma_1 \circ \big[\eta^{-1}(1, \cdots, \ell)\gamma_1^{-1} \circ \eta(\ell - 1, \ell, \ell - 2, \cdots, 1)\gamma_1\big] \\ = \eta(1, \cdots, \ell)\gamma_1 \circ \big[(\eta^{-1} \circ \eta)(\ell - 1, \ell - 2, \ell)(\gamma_1^{-1} \circ \gamma_1)\big] \\ = \eta(\ell - 2)\gamma_3\gamma_1 \in \langle \alpha; S_n \rangle\end{aligned}$$

where every circuit in the join $\gamma_2 = (\ell - 2)\gamma_3$ has length strictly less than ℓ, which finishes the proof. □

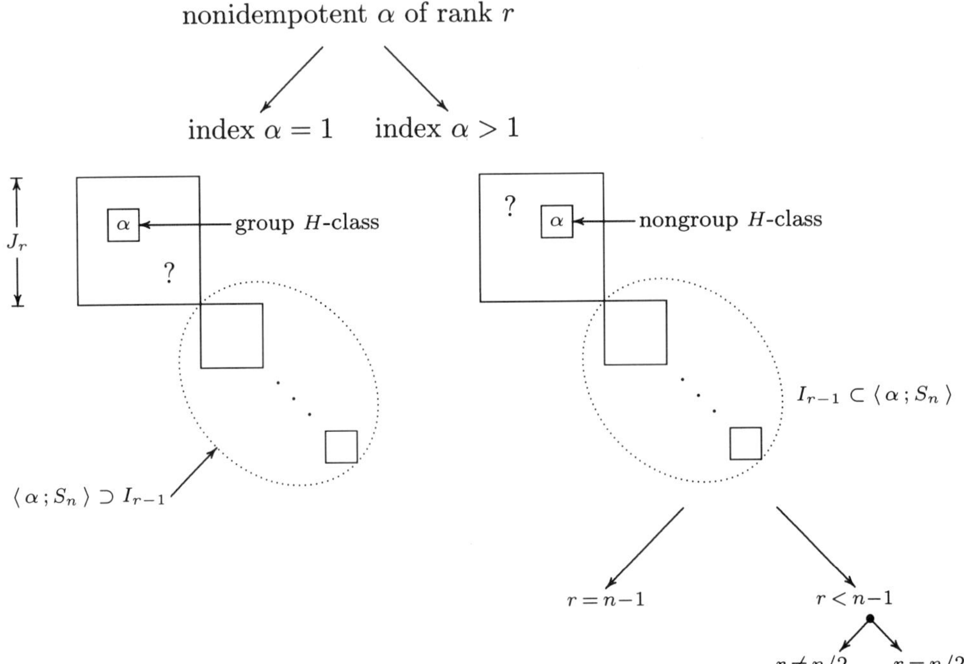

Figure 32.2. Cases for computing $\langle \alpha ; S_n \rangle$ when α is a nonidempotent.

Note that the charts α and β specified in Lemma 32.1 have the same rank because any two charts with the same nilpotent part have the same rank.

32.3 Proposition

Let $n \geq 5$ and suppose $\alpha \in C_n$ has index $m \geq 2$ and rank $n-1$. Then $\langle \alpha; S_n \rangle$ contains a rank $n-1$ chart whose path structure is either $(ijk]_{n-1}$ or $(jk]_{n-1}$.

Proof. Let $T = \langle \alpha; S_n \rangle$. Since m is the index of the rank $n-1$ chart α, we may suppose that $(1, 2, \cdots, m]\gamma$ is the decomposition of α. If $m = \ell \geq 3$, (1) of 28.2 shows that we may generate a 3-path of rank $n-1$ by multiplying $\alpha = (1, 2, \cdots, \ell]\gamma$ by the conjugate $(\ell - 1, \ell, \ell - 2, \cdots, 1]\gamma^{-1}$. If $m = 2$, we let $\alpha = (xy]\gamma$ and apply 32.1, obtaining $\beta = (xy]\gamma_1 \in T$ with each circuit in γ_1 having length at most 2. If each such circuit has length 1, then T contains a proper 2-path. Otherwise, since $n \geq 5$, we may suppose that either $\beta = (xy](12)(3)\lambda$ or $\beta = (xy](12)(34)\sigma$. But (7) of 28.2 reduces the latter case to the former, where (3) of 28.2 shows that T contains a rank $n-1$ proper 3-path. □

32.4 Lemma

Let $n \geq 4$ and suppose $T \subset C_n$ is S_n-normal. If $(ijk]_r \in T$ with $r < n-1$, then T contains a rank r proper 2-path $(ix]_r$.

Proof. Since $n \geq 4$ and $r < n-1$, there is a join ε of 1-paths such that $(ijk]_r = (ijk](x]\varepsilon$. But then

$$(ijk](x]\varepsilon \circ (kjx](i]\varepsilon = (ix](k](j)\varepsilon \in T,$$

which finishes the proof. □

32.5 Proposition

Let $n \geq 4$ and suppose $\alpha \in C_n$ has rank $r < n-1$, has index $m \geq 2$, but is not conjugate to $(12](34] \cdots (n-1, n]$. Then $\langle \alpha; S_n \rangle$ contains a proper 2-path $(jk]_r$ of rank r.

Proof. Let $T = \langle \alpha; S_n \rangle$. Since the index of α is m, we may suppose that $(1, 2, \cdots, m]\beta$ is the decomposition of α. If $m = \ell \geq 3$, then by (1) of 28.2,

$$(12 \cdots \ell]\beta \circ (\ell - 1, \ell, \ell - 2, \cdots, 1]\beta^{-1} = (\ell - 1, \ell - 2, \ell]_r \in \langle \alpha; S_n \rangle.$$

Applying 32.4, we conclude that $(ix]_r \in T$. If $m = 2$, then each proper path in the nilpotent part $(xy]\eta_1$ of $\alpha = (xy]\eta_1\gamma$ has length at most 2. In addition,

32.1 shows that there exists $\beta = (xy]\eta_1\beta_1 \in T$ such that each circuit in the join β_1 has length at most 2. Consequently, every path in $(xy]\eta_1\beta_1 = \beta$ has length at most 2. And since β and α have the same nilpotent part, β cannot be conjugate to $(12](34]\cdots(n-1,n]$, i.e., β cannot be a join of proper 2-paths. If $\eta_1\beta_1$ is a join of 1-paths, then we are finished because α and β have the same rank. Otherwise, the path decomposition of $\eta_1\beta_1$ contains some 2-paths. The possibilities, since $n \geq 4$ and $r < n-1$, reduce to T containing one of the following: (i) $(xy](12)(3]\lambda$, (ii) $(xy](12](3)\lambda$, (iii) $(xy](12](3]\lambda$, or (iv) $(xy](12](34)\sigma$. Then (i) gives the result by (4) of 28.2; (ii) by (5) of 28.2 and Lemma 32.4; (iii) by (6) of 28.2 and 32.4; and (iv) by (8) and then (4) of 28.2. □

Nonidempotents of Index 1

In contrast to the picture painted in the previous subsection, an S_n-normal semigroup T generated by charts with index 1 can meet a $\mathcal{J}$-class of C_n in a subset of index 1 charts.

32.6 Proposition

Let $n \geq 2$, let $0 \leq r \leq n$, and let $H_1, \ldots, H_{\binom{n}{r}}$ denote the maximal subgroups of C_n that are included in the $\mathcal{J}$-class J_r. If each chart in $A = \{\alpha_1, \ldots, \alpha_m\} \subset C_n$ has index 1 and rank r, then $T = \langle A; S_n\rangle$ meets J_r in a disjoint union $U = \vee_{1\leq i\leq\binom{n}{r}} G_i$ of conjugate normal subgroups $G_i \subset H_i$.

Proof. For each i, $1 \leq i \leq \binom{n}{r}$, let $G_i = T \cap H_i$. Then clearly G_i is a subgroup of H_i. To see that G_i is normal in H_i, let $\sigma \in G_i$ and suppose $\lambda = \gamma(j_1]\cdots(j_{n-r}] \in H_i$. Then $\alpha = \gamma(j_1)\cdots(j_{n-r}) \in S_n$ and $\lambda^{-1}\circ\sigma\circ\lambda = \alpha^{-1}\circ\sigma\circ\alpha \in T\cap H_i = G_i$. To see that each G_i is conjugate to each G_j, note that each H_i is conjugate to each H_j. So supposing $\alpha^{-1}H_i\alpha = H_j$ for some $\alpha \in S_n$, we have

$$\alpha^{-1}G_i\alpha = \alpha^{-1}(T\cap H_i)\alpha = (\alpha^{-1}T\alpha)\cap(\alpha^{-1}H_i\alpha) = T\cap H_j = G_j.$$

Since it is clear that $U \subset T\cap J_r$, we only need the reverse inclusion: If $\beta \in T\cap J_r$, then $\beta \in T = \langle A; S_n\rangle$ is a composition $\beta_1\circ\cdots\circ\beta_k$, where $\mathbf{d}\beta_1 = \mathbf{r}\beta_1 = \cdots = \mathbf{d}\beta_k = \mathbf{r}\beta_k$ because β and each of the α_i have index 1 and rank r. In other words, $\beta \in H_i$ for some i, implying $\beta \in T\cap H_i = G_i$. □

§33 Classification for $n \geq 5$

Recall that X_n denotes the X-semigroup on $n = 2k$ symbols, I_r the ideal of charts of rank $\leq r$, and E_m the semilattice of idempotents of rank $\leq m$.

33.1 Theorem

Let $n \geq 5$, let $1 \in S_n$ denote the identity, and let T be a subsemigroup of C_n. Then T is S_n-normal if and only if $T = S$ or $T = S \cup \{1\}$ where S is one of the following semigroups:

(1) *a normal subgroup G of S_n, or a union $G \cup I_r$, $0 \leq r \leq n-1$,*
(2) *the alternating semigroup A_n^c, or its ideal $A_n^c - A_n$,*
(3) *a union $E_m \cup I_r$, $0 \leq r \leq m \leq n-1$,*
(4) *a union $E_m \cup U \cup I_{r-1}$, $2 \leq r \leq m \leq n-1$, where $U = \vee_{1 \leq i \leq \binom{n}{r}} G_i$ is a disjoint union of nontrivial conjugate groups defined in* Proposition 32.6,
(5) *a union $E_m \cup X_n$, $n/2 \leq m \leq n-1$ (defined if and only if n is even).*

Proof. It is clear that if S is a semigroup listed in (1)–(5), then both S and $S \cup \{1\}$ are S_n-normal. Conversely, let T be S_n-normal.

Case 1. $T \cap S_n = \emptyset$.

Let $m = \max\{\,|\alpha| \mid \alpha \in T\}$ and suppose $r = \max\{\,|\alpha| \mid \alpha \in T - E_n\}$ (if all charts in T are idempotents, define $r = 0$). Consider the following subcases:

(1a) There exists $\alpha \in T \cap J_r$ with index ≥ 2.

First, assume α is not conjugate to $(12](34]\ldots(n-1,n]$. If $r < n-1$, then 32.5 and (30.1) show that $I_r \subset T$, implying $T = E_m \cup I_r$. If $r = n-1$, then 32.3 with (30.1) and (30.2) show that $A_n^c - A_n \subset T$. Thus, $T = A_n^c - A_n$ if every rank $n-1$ $a \in T$ is even. Otherwise, T contains a rank $n-1$ chart a that is odd. Then, by (30.1) and the fact that each chart of rank $n-1$ is either even or odd but not both, since α has rank $n-1$, we have $\alpha = (i_1, j_1] \circ (i_2, i_1] \circ (i_3, i_2] \circ \cdots \circ (i_k, i_{k-1}]$ where k is odd and $\mathbf{d}\alpha = N - \{j_1\}$. But then $\beta = (i_2, i_1] \circ (i_3, i_2] \circ \cdots \circ (i_k, i_{k-1}]$ is even and therefore in T. As a consequence, $\alpha \circ \beta^{-1} = (i_1, j_1] \in T$, showing, by (30.1), that $T = I_{n-1}$. Second, suppose $n = 2k$, $r = k$, and α together with every chart in T of rank r and index ≥ 2 is conjugate to $(12](34]\ldots(n-1,n]$. Then from §29, $T = E_m \cup X_n$.

(1b) All charts in $T \cap J_r$ have index 1 and some $\alpha \in T \cap J_r$ has period ≥ 2.

Then from 31.2 and 32.6, we have $I_{r-1} \subset T$ and $T \cap J_r = U$, implying $T = E_m \cup U \cup I_{r-1}$.

(1c) Every chart in T is an idempotent.

In this case, 31.1 shows $T = E_m$.

Case 2. $T \cap S_n \neq \emptyset$.

Let $G = T \cap S_n$ and note that G is a normal subgroup of S_n. Moreover, since $n \geq 5$, G is $\{1\}$, A_n, or S_n. Now consider the following subcases:

(2a) $T \subset S_n$.

In this case, $T = G$ is a normal subgroup of S_n.

(2b) $T \not\subset S_n$ and $G = \{1\}$.

Note that $T^- = T - S_n$ is an S_n-normal semigroup. By (1a), (1b), and (1c), the semigroup T^- is either the ideal $A_n^c - A_n$ in (2) or one of the semigroups defined in (3), (4), or (5). As a result, $T = T^- \cup \{1\}$.

(2c) $T \not\subset S_n$ and $G \neq \{1\}$.

In this case, note that $(123) \in T$. Moreover, if $r = \max\{\,|\alpha| \mid \alpha \in T^- = T - S_n\,\}$, then $E_r \subset T$. If $r = 0$, then $T = G \cup I_0$. Otherwise, if $1 \leq r \leq n-2$, upon multiplying (123) by a suitable rank r idempotent (having neither 2 nor 3 in its domain), we obtain $(12]_r \in T$. Applying (30.1), we have $T = G \cup I_r$. Similarly, if $r = n-1$ and $G = A_n$, then $(123]_{n-1} = 1_{N-\{3\}} \circ (123) \in T$. In this case, (30.2) implies $A_n^c - A_n \subset T^- = T - S_n$; and then the proof of (1a) above yields $T^- = A_n^c - A_n$ or $T^- = I_{n-1}$, implying $T = A_n^c$ or $T = A_n \cup I_{n-1}$. Finally, if $r = n-1$ and $G = S_n$, then $(12]_{n-1} = 1_{N-\{2\}} \circ (12) \in T$. And in this case, (30.1) shows $T = S_n \cup I_{n-1}$. □

§34 Classification for $n < 5$

For $n = 1$, each nonempty subset, I_0, S_1, and $I_0 \cup S_1$ of C_1 is an S_1-normal semigroup. For $n = 2$, we have 10 S_2-normal semigroups (Figure 28.1). For $n = 3$, we have 26 S_3-normal semigroups (Figure 34.1). This can be checked directly or deduced from an argument similar to the proof of Theorem 33.1.

For $n = 4$, we add to those kinds of semigroups listed in 33.1 certain semigroups related to the 61-*semigroup* (Table 26.2 and Figure 39.5):

$$T_{61} = \langle\ \{(12)(34), (12)(34]\};\ S_4\ \rangle.$$

As its subscript indicates, T_{61} contains 61 charts. From the proof of 26.1, $\alpha \in T_{61}$ if and only if α is a restriction of one of the four permutations in the Klein 4-group $V = \{(1), (12)(34), (13)(24), (14)(23)\}$. As a consequence, since $(12](34]$ is a restriction of $(12)(34) \in V$, the semigroup $X_4 \subset T_{61}$. It then follows that $T_{61} \cap I_2 = X_4$. As for rank 3, since each 3-subset of $N = \{1, 2, 3, 4\}$ has exactly two elements in common with every other 3-subset of N, each J_3 $\mathcal{H}$-class that is not a group contains a chart $(xy)(uv]$ conjugate to $(12)(34]$. It follows that $T_{61} \cap J_3$ contains exactly one chart in each of the 16 J_3 $\mathcal{H}$-classes. At rank 4 we have $T_{61} \cap S_4 = V$.

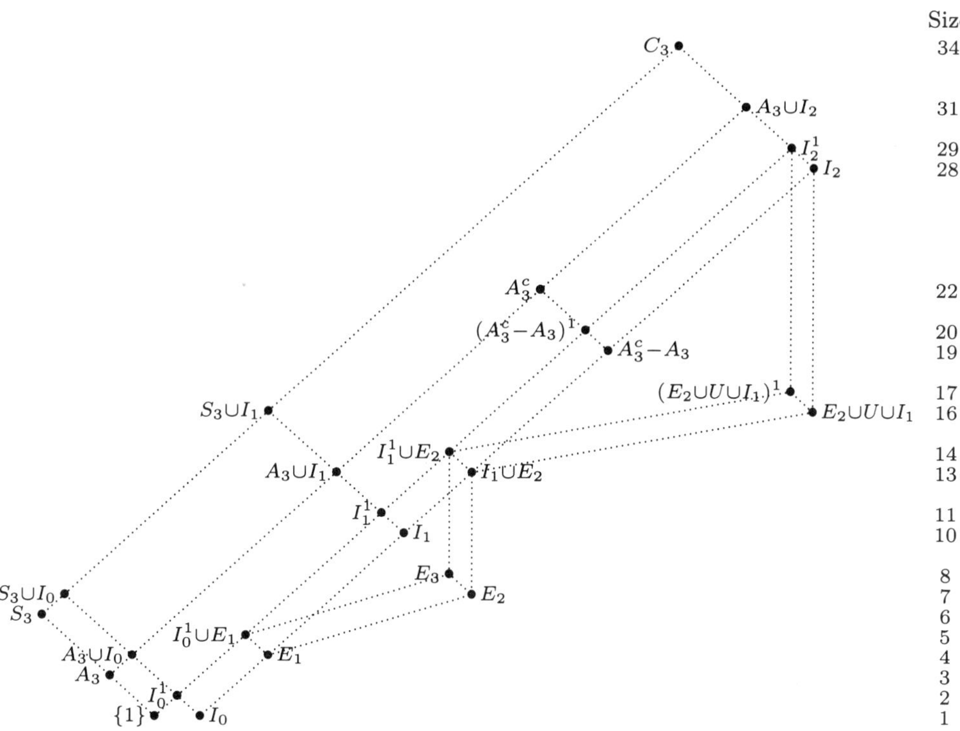

Figure 34.1. Lattice of S_3-normal semigroups.

Turning to subsemigroups of T_{61}, we define the 57-*semigroup*

$$T_{57} = T_{61} \cap I_3 = \langle\ \alpha^{-1}(12)(34]\alpha \mid \alpha \in S_4\ \rangle.$$

This semigroup is S_4-normal and its name is derived from its order ($61 - 4 = 57$). From T_{57} we derive the 105-*semigroup* $T_{105} = T_{57} \cup I_2$, another S_4-normal semigroup.

34.2 Theorem

Let V be the Klein 4-group. Then T_{61}, T_{57}, $\{1\} \cup T_{57}$, T_{105}, $\{1\} \cup T_{105}$, $V \cup T_{105}$, $V \cup (A_4^c - A_4)$, and $V \cup X_4$ are the only S_4-normal semigroups not specified in Theorem 33.1.

Proof. Let T be S_4-normal and suppose i, j, k, and t are distinct members of $\{1, 2, 3, 4\}$. Letting U denote the appropriate disjoint union of groups as defined in Theorem 34.1, we analyze the possible subsets of T at each rank.

We begin with ranks 4 and 3:

rank 4 (index = 1)	rank 3 (index > 1)
$(ij) \in T \Rightarrow S_4 \subset T$	$(i)(j)(kt] \in T \Rightarrow I_3 \subset T$
$(ijk) \in T \Rightarrow A_4 \subset T$	$(i)(jkt] \in T \Rightarrow A_4^c - A_4 \subset T$
$(ij)(kt) \in T \Rightarrow V \subset T$	$(ijkt] \in T \Rightarrow$ (by (1) of 28.2) $A_4^c - A_4 \subset T$
$(i)(j)(k)(t) \in T \Rightarrow \{1\} \subset T$	$(ij)(kt] \in T \Rightarrow T_{57} \subset T$ or $T_{105} \subset T$

rank 3 (index = 1)
$\alpha \in T \Rightarrow E_3 \subset T$ or $U \subset T$

Next, in the rank 2 case $(i](jkt] \in T$ below, it helps to observe that the multiplication $(t](ijk] \circ (i](kjt] = (j)(it](k]$.

rank 2 (index > 1)	rank 1 (index > 1)
$(i)(jk](t] \in T \Rightarrow I_2 \subset T$	$(ij](k](t] \in T \Rightarrow I_1 \subset T$
$(i](jkt] \in T \Rightarrow I_2 \subset T$	
$(ij](kt] \in T \Rightarrow X_4 \subset T$	

rank 2 (index = 1)	rank 1 (index = 1)
$(ij)(k](t] \in T \Rightarrow U \subset T$	$(i)(j](k](t] \in T \Rightarrow E_1 \subset T$

Because of these inclusions, the result follows. □

§35 Comments

Chapter III of Petrich's text [1] is a good reference for the general theory of congruences of inverse semigroups and the role played by normal semigroups. The term S_n-normal derives from the more general term $\mathcal{G}_X$-normal: If $\mathcal{G}_X$ denotes the symmetric group of permutations on a set X, then a semigroup S of transformations (total or partial) on X is called $\mathcal{G}_X$*-normal* if $\alpha^{-1}S\alpha \subset S$ for all $\alpha \in \mathcal{G}_X$. (In our case, $X = N$.)

For X finite, the $\mathcal{G}_X$-normal semigroups of total transformations were classified in 1976 by J. S. V. Symons [1]. A similar "to appear" result for partial transformations was announced in 1980 by Sullivan [3, Ref. 8]. For X infinite, $\mathcal{G}_X$-normal semigroups of one-one transformations were studied in 1990 and 1991 (I. Levi and W. Williams [1] and [2], and Levi [1]).

The classification of the S_n-normal semigroups did not appear until 1995 when, along with the classification, the X semigroups, the T_{57} semigroup, and the T_{105} semigroup were first introduced (Lipscomb and Konieczny [1]).

In fact, except for equation (30.1), the statements and proofs of the theorems presented here were first introduced in that paper. As for equation (30.1), it was first proved in 1987 by Sullivan [1, Lemma 2].

CHAPTER 8

Normal Semigroups and Congruences

Just as the alternating group A_n is normal in S_n, the alternating semigroup A_n^c is normal in C_n. But unlike the group case, A_n^c cannot (when $n \geq 3$) be the kernel of a congruence on C_n. Similarly, the $E_n \cup X_n$ semigroups ($n \geq 4$) are an abnormal kind of normal semigroup.

In general, a subsemigroup K of C_n is the kernel of a congruence on C_n if and only if K is both normal and has a partition $K = I \cup G_1 \cup G_2 \cup \ldots \cup G_t$ ($t \geq 0$) where I is an ideal and each G_i is a group. As a matter of fact, for each such K, the ideal I determines the partition and there results a one-one correspondence between such pairs (K, I) and the congruences on C_n that have K as their kernel.

The C_2 and C_3 cases expose the pattern (§36), while the general case depends on decomposing the semilattice E of idempotents in C_n (§37). In §38, we learn that "kernels" are characterized as normal semigroups that have a decomposition $I \cup G_1 \cup \cdots \cup G_t$ as described above. In §39, we apply the classification of S_n-normal semigroups (Chapter 7) to show that the only normal and nonkernel semigroups in C_n ($n \geq 5$) are the alternating semigroups A_n^c and the semigroups $E_n \cup X_n$ (which exist only when n is even). For $n = 2$, each normal semigroup is a kernel; for $n = 3$, each normal semigroup except A_3^c is a kernel; and, for $n = 4$, each normal semigroup except A_4^c, $E_4 \cup X_4$, and several derivatives of T_{61} is a kernel. From the results in §39, we derive a formula for the number of C_n congruences (§40).

§36 Examples[1]

Letting $n = 2$ and using "1" to denote the identity (1)(2) of C_2, we see in Figure 36.1 the three full self-conjugate subsemigroups of C_2 — we see the S_2-normal semigroups in Figure 28.1 that contain all idempotents of C_2. Observe that these three semigroups, $E_2 \subset I_1^1 \subset C_2$, yield to pairwise disjoint decompositions of the form "ideal" or "ideal $\cup$ group $\cup \cdots \cup$ group." And further inspection shows that only one of these three, namely C_2 itself, has two such decompositions — "I_2" and "$I_1 \cup S_2$." In total, we have four decompositions of the three normal semigroups. These four decompositions correspond one-one with the four congruences on C_2. In fact, letting $\Delta T =$

[1] The reader may want to return to these examples after reading §38.

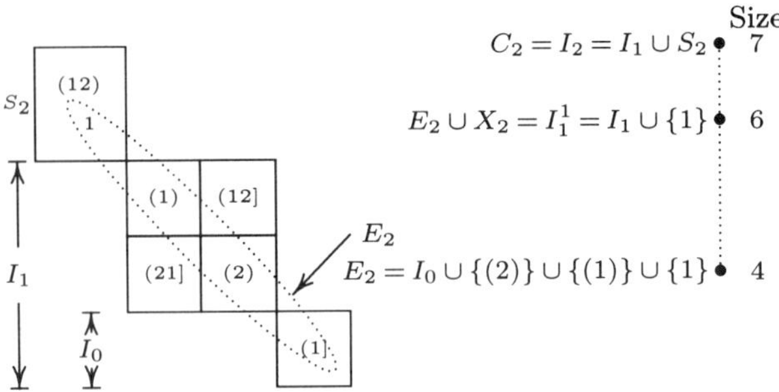

Figure 36.1. Egg-box of C_2 and lattice of normal subsemigroups.

$\{(t,t) \mid t \in T\}$, we can illustrate this correspondence:

$$\begin{array}{lclcl} (C_2, I_2) & \leftrightarrow & I_2 & \leftrightarrow & I_2 \times I_2 \\ (C_2, I_1) & \leftrightarrow & I_1 \cup S_2 & \leftrightarrow & (I_1 \times I_1) \cup (S_2 \times S_2) \\ (I_1^1, I_1) & \leftrightarrow & I_1 \cup \{1\} & \leftrightarrow & (I_1 \times I_1) \cup \Delta S_2 \\ (E_2, I_0) & \leftrightarrow & I_0 \cup \{(2)\} \cup \{(1)\} \cup \{1\} & \leftrightarrow & (I_0 \times I_0) \cup \Delta(C_2 - I_0). \end{array}$$

Turning to $n = 3$ and now using "1" to denote the identity $(1)(2)(3) \in C_3$, we may use the lattice in Figure 34.1 to find the seven normal semigroups in C_3, which are those S_3-normal semigroups that contain all idempotents of C_3. Among these seven, the six pictured in Figure 36.2 yield to pairwise disjoint decompositions of the form "ideal" or "ideal $\cup$ group $\cup \cdots \cup$ group," A_3^c being the only outlier.

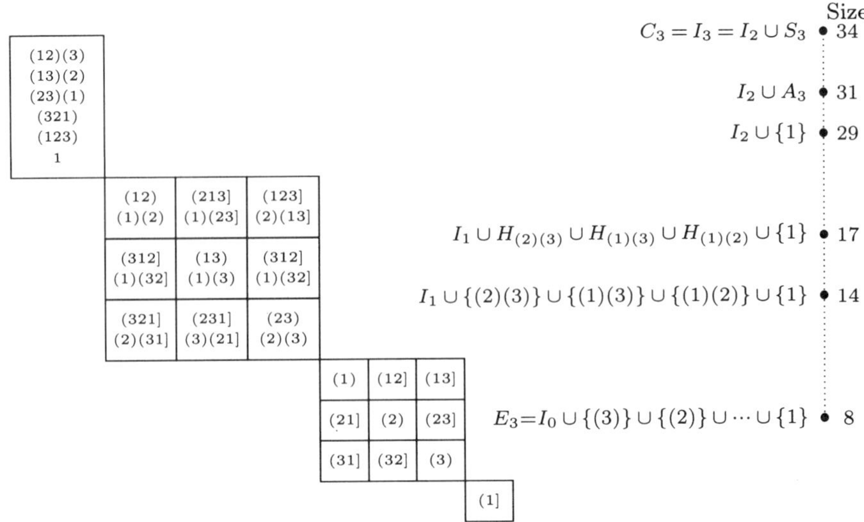

Figure 36.2. Egg-box of C_3 and lattice of kernel-normal subsemigroups.

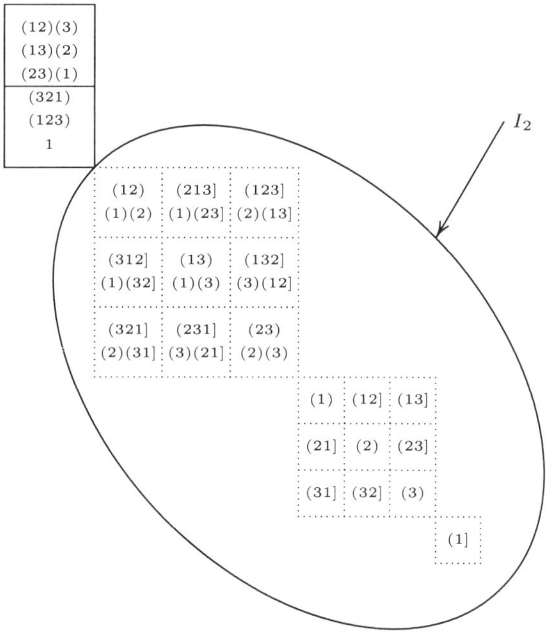

Figure 36.3. Quotient set C_3/ρ relative to the egg-box of C_3.

As in the $n = 2$ case, exactly one normal subsemigroup of C_3, namely C_3 itself, exhibits more than one such decomposition. By what follows, C_3 has precisely seven congruences, which are in one-one correspondence with the decompositions displayed in Figure 36.2.

Before going on to formal arguments, however, it is instructive to overlay the egg-box picture of C_3 with the partition of C_3 induced by one of its congruences. For example, consider the congruence ρ induced by the decomposition $I_2 \cup A_3$. The ideal I_2 has the effect of "removing" those edges of the C_3 egg-box that bound the elements of I_2, while the group A_3 has the effect of "adding" one edge to the "rank 3" egg-box (Figure 36.3).

§37 Congruences on Finite Semilattices

Recall that a *semilattice* E is a commutative semigroup of idempotents, that the elements of each such E have a *natural ordering* ($e \leq f$ when $ef = e$), and that together $(E, \leq)$ is a partially ordered set with greatest lower bound function "glb" given by $\text{glb}\{e, f\} = ef$. Moreover, recall that it works in reverse. Given a partially ordered set $(S, \preceq)$ with a greatest lower bound function "glb", we get a semilattice $E = S$ whose multiplication $E \times E \to E$ is given by $ef = \text{glb}\{e, f\}$ and whose natural ordering $\leq$ is $\preceq$. (Our model for a semilattice is the subsemigroup E of idempotents in C_n.)

When $f \le e$, we sometimes say that e *follows* f (e is above f) or that f *precedes* e (f is below e), and we may write $e \ge f$. Each finite semilattice $E = \{e, f, \dots, h\}$ has a *minimum element* $ef \cdots h$ that precedes every element of E. Whenever e follows f and $f \le h \le e$ implies $f = h$ or $h = e$, we shall say that e *covers* f . If e does not cover f but merely follows f, then, since E is finite, there exist $f_1, \dots, f_k \in E$ such that $f \le f_1 \le \cdots \le f_k \le e$ and f_1 covers f, f_2 covers f_1, ..., e covers f_k.

For an ideal I in E, we shall need the set A_I of elements that are "above and adjacent"to I — we call

$$A_I = \{ e \in E - I \,|\, e \text{ covers some } f \in I \}$$

the *cover* of I. Note that $A_I \ne \emptyset \Leftrightarrow I \ne E$ because every ideal I contains the first element e_1 of E and each $e \in E - I$ follows e_1. We also need to "reduce" A_I by removing those members that follow members of A_I, i.e., for a given ideal I in E, we define its *reduced cover*

$$R_I = A_I - \{ h \in A_I \,|\, h \text{ follows some other } e \in A_I \}.$$

A reduced cover R_I is not necessarily a cover (Figure 37.1).

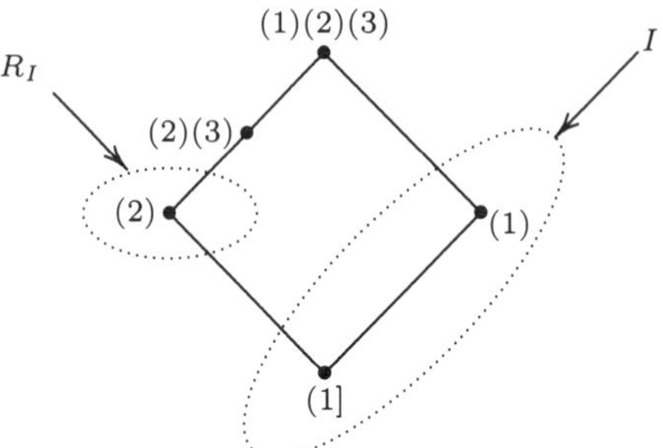

Figure 37.1. Reduced cover R_I that is not a cover of I.

The key relation between R_I and I is that every pair of distinct members e and f of R_I has their product ef in I. Reduced covers and the following proposition are needed to prove 37.3.

37.2 Proposition

A nonempty subset I of a semilattice E is an ideal in E if and only if $f \le e \in I \Rightarrow f \in I$.

Proof. ($\Rightarrow$) Let I be an ideal in E and suppose $f \le e \in I$. Then $f = fe$ and, since I is an ideal, $f = fe \in I$. ($\Leftarrow$) Suppose I is a subset of E such that $f \le e \in I \Rightarrow f \in I$. Then for $h \in E$ and $e \in I$, we have $he \le e \in I$, implying $he \in I$. □

It is clear from 37.2 that the natural ordering of a semilattice determines its Rees congruences — a congruence τ on E is a *Rees congruence* when there exists an ideal I in E such that $\tau = (I \times I) \cup \{(e,e) \mid e \in E - I\}$. If $I = E$, then we have the *trivial congruence* $E \times E$. All other congruences, Rees or otherwise, are called *nontrivial.*

(To quickly gain an understanding of the following proposition, the reader may find Figure 37.4 useful.)

37.3 Proposition

Let τ be a nontrivial congruence on the finite semilattice E whose first element is e_1. Then E has a partition

$$E = K_0 \cup K_{11} \cup K_{12} \cup \cdots \cup K_{1r_1} \cup \cdots \cup K_{m1} \cup K_{m2} \cup \cdots \cup K_{mr_m}$$

where

(1) *the set $K_0 = \{ f \in E \mid f\tau e_1 \}$ is an ideal of E,*
(2) *for each $i = 1, \ldots, m$, the set $K_i = K_{i-1} \cup K_{i1} \cup \cdots \cup K_{ir_i}$ is an ideal of E, and*
(3) *if $R_i = \{e_{i1}, \ldots, e_{ir_i}\}$ denotes the reduced cover of the ideal K_{i-1}, then for $j = 1, \ldots, r_i$, the set $K_{ij} = \{ f \in E \mid e_{ij} \leq f$ and $e_{ik} \not\leq f$ for all $k \neq j \}$ is a subsemigroup of E.*

Moreover, for each τ-class $e\tau = \{ f \in E \mid f\tau e \} \in E/\tau$, either $e\tau = K_0$ or $e\tau \subset K_{ij}$ for some i and some j. And in particular, if $E = E_n$ is the semilattice of idempotents in C_n and K_0 consists of all idempotents of rank $\leq r$, then each K_{ij} is a trivial group whose lone member ε_{ij} is an idempotent of rank $r+i$ and each $r_i = \binom{n}{r+i}$ is the number of idempotents of rank $r+i$. In this particular case, then, τ is a Rees congruence on E_n.

Proof. The proof is by induction on the subscript i of K_i. To see that K_0 is an ideal, note that if $h \in E$ and $f \in K_0$, then $f\tau e_1$ yields $fh\tau e_1 h = e_1$. So K_0 is an ideal. Since τ is nontrivial and E is finite, the reduced cover $R_1 = \{e_{11}, \ldots, e_{1r_1}\}$ of K_0 is nonempty. Each K_{1j} is therefore well defined, and, since $e_{1j} \in K_{1j}$, each $K_{1j} \neq \emptyset$. To see that K_{1j} is a semigroup, let $f, h \in K_{1j}$. Then e_{1j} is a lower bound of both f and h, yielding $e_{1j} \leq fh$ for only j. In other words, $fh \in K_{1j}$. Thus, (3) above holds for $i = 1$. To see that (2) holds for $i = 1$, i.e., that K_1 is an ideal, we use 37.2. Let $e \leq f \in K_1$. If $f \in K_0$, then, since K_0 is an ideal, 37.2 shows that $e \in K_0 \subset K_1$. If $f \notin K_0$, then $f \in K_{1j}$ for some j, implying $e_{1j} \leq f$ while $e_{1k} \not\leq f$ for all $k \neq j$. Now if $e_{1k} \leq e$ for some k, then $e_{1k} \leq e \leq f$; and then e can follow only e_{1j}, yielding $e \in K_{1j}$. If e does not follow any e_{ik}, then, since R_1 is the reduced cover of K_0, we have $e \in K_0 \subset K_1$. Moreover, since it is clear that the representation for K_1 is a partition, we are finished with the first step of our induction. The proof of the inductive step, however, is similar to the proof

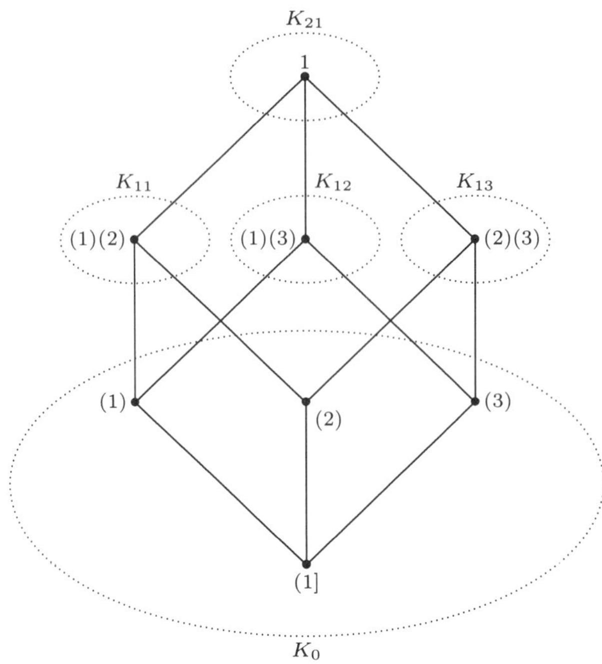

Figure 37.4. Semilattice $E_3 = K_0 \cup K_{11} \cup K_{12} \cup K_{13} \cup K_{21}$ in C_3.

for first step and is left to the reader. It therefore follows, since E is finite, that we may decompose E as indicated. That the resulting decomposition is a partition follows from our inductive construction and the proper nesting (proper inclusion chain) of the ideals $K_0 \subset K_1 \subset \cdots \subset K_m = E$.

To see that each τ-class $e\tau \in E/\tau$ is either K_0 or a subset of some K_{ij}, suppose that K_{i-1} is a union of τ-classes and let $e \in K_{ij}$. Then $e \geq e_{ij}$ and $e \not\geq e_{ik}$ for every $k \neq j$. Suppose $e\tau h$ where $h \notin K_{i-1} \cup K_{ij}$. Then $h \geq e_{ik}$ for some $k \neq j$ and $ee_{ik}\tau he_{ik} = e_{ik}$. Since both e and e_{ik} have ee_{ik} as a lower bound, we must have $ee_{ik} < e_{ik}$. But then $ee_{ik} \in K_{i-1}$ and $ee_{ik}\tau e_{ik} \notin K_{i-1}$ contradicts the fact that K_{i-1} is a union of τ classes. Thus, $h \in K_{i-1} \cup K_{ij}$ and we are finished with the proof that each τ-class $e\tau \in E/\tau$ is either K_0 or a subset of K_{ij}.

To see that each K_{ij} must be a trivial group when $E = E_n$ is the set of idempotents in C_n and K_0 consists of all idempotents whose rank is $\leq r$, we first observe that the reduced cover of K_0 is the set of idempotents of rank $r+1$ and that $r_1 = \binom{n}{r+1}$ (Figure 37.4). Then, since each idempotent that follows a rank $r+1$ idempotent follows at least two idempotents of rank $r+1$, we see from the definition of K_{1j} that each $K_{1j} = \{\varepsilon_{1j}\}$ is a trivial group. By induction on the subscript i of K_i, we find that each K_{ij} is the trivial group $\{\varepsilon_{ij}\}$. Finally, for each τ-class $\varepsilon_{ij}\tau \in E/\tau$, the inclusion $\varepsilon_{ij}\tau \subset K_{ij} = \{\varepsilon_{ij}\}$ shows that τ is indeed a Rees congruence on E_n. □

§38 Kernels of Congruences

Let E_n be the semilattice of idempotents in C_n and suppose ρ is a congruence on C_n. Recall that the *trace* "tr ρ" of ρ is the congruence $\tau = \{ (\varepsilon, \varepsilon') \mid \varepsilon, \varepsilon' \in E_n \text{ and } \varepsilon\,\rho\,\varepsilon' \}$ on E_n, and that the *kernel* "ker ρ" of ρ is $\{ \alpha \in C_n \mid \alpha\,\rho\,\varepsilon \text{ for some } \varepsilon \in E_n \}$.

38.1 Lemma

Let ρ be a congruence on C_n, let E_n be the semilattice of idempotents in C_n, let $I = [0]_\rho = \{\alpha \in C_n \mid \alpha\,\rho\,0\}$, and, for each $0 \le r \le n$, let $I_r \subset C_n$ be the ideal of all charts of rank $\le r$. Then

(*i*) *for some r, we have $I = I_r$,*
(*ii*) *if $\varepsilon \in E_n - I_r$, then $G(\varepsilon) = \{ \beta \in C_n \mid \beta\,\rho\,\varepsilon \}$ is a subgroup of C_n, and*
(*iii*) *if $K = \ker\rho = \{ \alpha \in C_n \mid \alpha\,\rho\,\varepsilon$ for some $\varepsilon \in E_n \}$, then K is a normal semigroup in C_n and has a partition $K = I \cup G_1 \cup G_2 \cup \cdots \cup G_t$ $(t \ge 0)$ where I is the ideal specified in (i), and where each G_i is a group $G(\varepsilon)$ as specified in (ii).*

Proof. For (*i*), show that I is an ideal of C_n, and that the only ideals of C_n are $I_0, I_1, \dots, I_n$. For (*ii*), show that each $G(\varepsilon)$ is necessarily an inverse semigroup, and, in particular, a group whenever it contains only one idempotent. Then consider $\tau = \operatorname{tr}\rho$ on E_n and apply 37.3. The proof of (*iii*) requires several observations. First, K is normal: It is self-conjugate, since $\beta\,\rho\,\varepsilon \in E_n$ and $\alpha \in C_n$ imply $(\alpha^{-1}\beta\alpha)\,\rho\,(\alpha^{-1}\varepsilon\alpha) \in E_n$, and it is full, since $E_n \subset K$. Second, let $\varepsilon_1, \dots, \varepsilon_t$ be the idempotents in $C_n - I$. Then, setting $G_i = G(\varepsilon_i)$, we clearly have $K = I \cup G_1 \cup \cdots \cup G_t$, where each G_i is a group as specified in (*ii*). □

Because of (*iii*) in 38.1, we shall say that a normal subsemigroup K of C_n is *kernel normal* if it has a partition $K = I \cup G_1 \cup \cdots \cup G_t$ $(t \ge 0)$ where I is an ideal in C_n and each G_i is a group.

38.2 Lemma

Let $0 < r \le n$. If K is a kernel-normal semigroup in C_n with partition $K = I \cup G_1 \cup \cdots \cup G_t$ $(t \ge 0)$ where $I = I_{r-1}$ is an ideal in C_n, then each group G_i whose elements are of rank $> r$ is a trivial group.

Proof. Suppose the idempotent ε of G_i has rank $> r$ and that $\varepsilon \neq \alpha \in G_i$.

Then $x\alpha = y \neq x$ for some $x, y \in N = \{1, \dots, n\}$. Choosing distinct idempotents $\varepsilon_x, \varepsilon_y \in C_n$ of rank $|\varepsilon| - 1 \geq r$ such that $\varepsilon = (x)\varepsilon_x = (y)\varepsilon_y$, we calculate that $\varepsilon_x \circ \alpha = \alpha \circ \varepsilon_y$, and, since $\varepsilon \,\rho\, \alpha$, that $\varepsilon_x = (\varepsilon_x \circ \varepsilon)\,\rho\,(\varepsilon_x \circ \alpha) = (\alpha \circ \varepsilon_y)\,\rho\,(\varepsilon \circ \varepsilon_y) = \varepsilon_y$, contradicting the fact that each G_j is a group. □

Item (*iii*) of 38.1 defines a map $\rho \mapsto (\ker\rho, [0]_\rho)\}$ from the set $\{\rho\}$ of congruences on C_n into the set $\{(K, I)\}$ of ordered pairs (K, I) where K is kernel normal and $I \subset K$ is an ideal in C_n. Our next theorem shows that this map has an inverse $(K, I) \mapsto \rho_{(K,I)}$. In other words, these ordered pairs correspond one-one with the congruences on C_n.

38.3 Theorem

If K is a kernel-normal semigroup in C_n and $I \subset K$ is an ideal in C_n, then $\rho_{(K,I)} \subset C_n \times C_n$ given by

$$\alpha\rho_{(K,I)}\beta \;\Leftrightarrow\; \alpha, \beta \in I \quad or \quad (\alpha \mathcal{H} \beta \quad and \quad \alpha\beta^{-1} \in K)$$

is a congruence on C_n with $K = \ker \rho_{(K,I)}$ and I equal to the $\rho_{(K,I)}$-class $[0]$. In reverse, if ρ is any congruence on C_n, then

$$\rho = \rho_{(\ker \rho, I)}$$

where $\ker \rho = K$ is a kernel-normal semigroup in C_n with ρ-class $[0] = I \subset K$ an ideal in C_n.

Proof. We begin by showing that $\rho_{(K,I)}$ is an equivalence relation. Now $\rho_{(K,I)}$ is reflexive since K is full, and it is symmetric since K is an inverse semigroup. To see that $\rho_{(K,I)}$ is transitive, suppose $\alpha\,\rho_{(K,I)}\,\beta\,\rho_{(K,I)}\,\gamma$. If $\beta \in I$, then $\alpha, \gamma \in I$ and $\alpha\,\rho_{(K,I)}\,\gamma$. Otherwise, $\alpha\,\mathcal{H}\,\beta$ with $\alpha\beta^{-1} \in K$, and $\beta\,\mathcal{H}\,\gamma$ with $\beta\gamma^{-1} \in K$. Then $\alpha\,\mathcal{H}\,\gamma$ and $\alpha\gamma^{-1} = \alpha\alpha^{-1}\alpha\gamma^{-1} = (\alpha\beta^{-1})(\beta\alpha^{-1}) \in K$ (since $\alpha\,\mathcal{H}\,\beta$ implies $\alpha^{-1}\alpha = \beta^{-1}\beta$), which shows $\alpha\,\rho_{(K,I)}\,\gamma$. Thus, $\rho_{(K,I)}$ is an equivalence relation. To see that it is also a congruence, we begin by showing that it is a right congruence. So let $\alpha\,\rho_{(K,I)}\,\beta$ and suppose that $\gamma \in C_n$. We show $\alpha\gamma\,\rho_{(K,I)}\,\beta\gamma$. If $\alpha\gamma \in I$, then $\beta\gamma$ is also in I, since $\beta\gamma = \beta\beta^{-1}\beta\gamma = \beta\alpha^{-1}(\alpha\gamma) \in I$ (because $\beta^{-1}\beta = \alpha^{-1}\alpha$). Otherwise, for r given by $I = I_{r-1}$, we have $|\alpha\gamma| \geq r$, and consequently $\alpha\,\mathcal{H}\,\beta$ with $\alpha\beta^{-1} \in K$. Moreover, if $|\alpha\gamma| > r$, then $|\alpha| > r$, and therefore, 38.2 implies $\alpha = \beta$. In fact, if either $|\alpha|$ or $|\beta| > r$, then $\alpha = \beta$ and we are finished. So we are left with the case $|\alpha\gamma| = |\alpha| = |\beta| = |\beta\gamma| = r$. First, since $\mathcal{L}$ is a right congruence, $\alpha\,\mathcal{L}\,\beta$ implies $\alpha\gamma\,\mathcal{L}\,\beta\gamma$. Second, $\mathbf{d}(\alpha\gamma) \subset \mathbf{d}\alpha$ and $|\alpha\gamma| = |\alpha|$ imply $\mathbf{d}(\alpha\gamma) = \mathbf{d}\alpha$. And since similarly $\mathbf{d}(\beta\gamma) = \mathbf{d}\beta$, we have $\mathbf{d}(\alpha\gamma) = \mathbf{d}(\beta\gamma)$, showing $\alpha\gamma\,\mathcal{R}\,\beta\gamma$. It follows that $\alpha\gamma\,\mathcal{H}\,\beta\gamma$. Finally, since K is full,

$$(\alpha\gamma)(\beta\gamma)^{-1} = \alpha\gamma\gamma^{-1}\beta^{-1} = \alpha\alpha^{-1}\alpha\gamma\gamma^{-1}\beta^{-1}(\alpha\gamma\gamma^{-1}\alpha^{-1})\alpha\beta^{-1} \in K.$$

Since the proof that $\rho_{(K,I)}$ is a left congruence is similar, we conclude that $\rho_{(K,I)}$ is a congruence. To see that the $\rho_{(K,I)}$-class $[0] = I$, note that I itself is a $\rho_{(K,I)}$-class and that $0 \in I$. To show that $K = \ker \rho_{(K,I)}$, we use the fact that K is kernel normal. Indeed, for $\{\varepsilon_1, \ldots, \varepsilon_t\} = E_n - I$, we have $K - I = H(\varepsilon_1) \cup \cdots \cup H(\varepsilon_t)$ where $H(\varepsilon_i)$ is a group containing ε_i. But in addition, by *(iii)* of 38.1, we also have $\ker \rho_{(K,I)} - I = G(\varepsilon_1) \cup \cdots \cup G(\varepsilon_t)$ where each $G(\varepsilon_i)$ is a group containing ε_i. It therefore clearly suffices to show that $H(\varepsilon_i) = G(\varepsilon_i)$ for each i. But

$$\alpha \in G(\varepsilon_i) \;\Leftrightarrow\; \alpha\, \rho_{(K,I)}\, \varepsilon_i \;\Leftrightarrow\; (\alpha\, \mathcal{H}\, \varepsilon_i \text{ and } \alpha = \alpha\varepsilon_i^{-1} \in K) \;\Leftrightarrow\; \alpha \in H(\varepsilon_i).$$

Next, suppose ρ is any congruence on C_n. Then 38.1 shows that $K = \ker \rho$ is kernel normal with the ρ-class $[0] = I \subset K$ as an ideal in C_n. To see that $\rho \subset \rho_{(K,I)}$, suppose $\alpha\, \rho\, \beta$. If $\alpha \in I$, then $0\, \rho\, \alpha\, \rho\, \beta$ implies $\beta \in I$, showing $\alpha\, \rho_{(K,I)}\, \beta$. If $\alpha \notin I$, then, since $\alpha\, \rho\, \beta$ implies $\alpha^{-1}\, \rho\, \beta^{-1}$, we have $\alpha\alpha^{-1}\, \rho\, \alpha\beta^{-1}\, \rho\, \beta\beta^{-1}$ and $\alpha^{-1}\alpha\, \rho\, \alpha^{-1}\beta\, \rho\, \beta^{-1}\beta$. These ρ equivalences not only show that $\alpha\alpha^{-1} = \beta\beta^{-1}$ and $\alpha^{-1}\alpha = \beta^{-1}\beta$ (since $\ker \rho$ is kernel normal and $\alpha \notin I$), but also that $\alpha\beta^{-1} \in \ker \rho = K$ (because $\alpha\alpha^{-1}$ is an idempotent). It follows that $\alpha\, \mathcal{H}\, \beta$ and, consequently, $\alpha\, \rho_{(K,I)}\, \beta$. To see the reverse inclusion, suppose $\alpha\, \rho_{(K,I)}\, \beta$. If $\alpha \in I$, then $\beta \in I$ and so $\alpha\, \rho\, 0\, \rho\, \beta$. If $\alpha \notin I$, then $\alpha\, \mathcal{H}\, \beta$ and $\alpha\beta^{-1} \in K - I$. Now $\alpha\, \mathcal{H}\, \beta$ gives $\alpha\alpha^{-1}\, \mathcal{H}\, \alpha\beta^{-1}$, while $\alpha\beta^{-1} \in K - I$ yields $\alpha\beta^{-1}\, \rho\, \varepsilon \in E_n - I$. By 38.1, both $\alpha\beta^{-1}$ and ε are in the same subgroup $G(\varepsilon)$ of K, showing that $\alpha\beta^{-1}\, \mathcal{H}\, \varepsilon$. It follows that $\varepsilon = \alpha\alpha^{-1}$ and that $\alpha\beta^{-1}\, \rho\, \alpha\alpha^{-1}$. And then we see that

$$\alpha = \alpha\alpha^{-1}\alpha = (\alpha\beta^{-1})\beta\, \rho\, (\alpha\alpha^{-1})\beta = \beta\beta^{-1}\beta = \beta,$$

since ρ and $\alpha\, \mathcal{H}\, \beta$ provide $\alpha\alpha^{-1} = \beta\beta^{-1}$ and $\alpha^{-1}\alpha = \beta^{-1}\beta$. □

§39 Kernels and Normal Semigroups

We specify all normal and kernel normal semigroups in C_n.

39.1 Proposition

A subsemigroup of C_n is normal if and only if it is full and S_n-normal.

Proof. The only real question is when K is full and $\alpha^{-1}K\alpha \subset K$ for every $\alpha \in S_n$. Then for $\beta \in C_n - S_n$, we show that $\beta^{-1}K\beta \subset K$. First, let $\varepsilon = \beta^{-1}\beta$ be the idempotent with domain $\mathbf{r}\beta$. Then choose a permutation $\gamma \in S_n$ such that $\gamma \circ \varepsilon = \beta$, e.g., as in §25, let $\gamma = \overline{\beta}$. Since K is full, it follows that $\beta^{-1}K\beta = (\gamma \circ \varepsilon)^{-1}K(\gamma \circ \varepsilon) = \varepsilon(\gamma^{-1}K\gamma)\varepsilon \subset \varepsilon K \varepsilon \subset K$. □

39.2 Proposition

Let $n \geq 5$, let X_n be the X-semigroup on $n = 2k$ symbols, let $I_r \subset C_n$ be the ideal of charts of rank $\leq r$, and let E_n be the semilattice of idempotents of C_n. Then K is a normal subsemigroup of C_n if and only if K is one of the following semigroups:

(1) *$I_{r-1} \cup U \cup E_n$, where $U = \vee_{1 \leq i \leq \binom{n}{r}} G_i$ is a disjoint union of conjugate groups defined in* Proposition 32.6*;*
(2) *the alternating semigroup A_n^c;*
(3) *$E_n \cup X_n$, which is only defined when n is even.*

Proposition 39.2 follows from 39.1 and 33.1; and the next two corollaries result from checking the list in 39.2 for kernel-normal semigroups.

39.3 Corollary

Let $n \geq 5$. Then the alternating semigroup A_n^c and the semigroup $E_n \cup X_n$ (when n is even) are the only normal semigroups in C_n that are not kernels of any congruences on C_n.

39.4 Corollary

Let $n \geq 5$. Then a normal semigroup K in C_n is the kernel of a congruence on C_n if and only if there exists an r, $0 < r \leq n$, such that $K = I_{r-1} \cup U \cup E_n$, where $U = \vee_{1 \leq i \leq \binom{n}{r}} G_i$ is a disjoint union of conjugate groups defined in Proposition 32.6.

Turning to C_n for $n < 5$, we start with C_2, where every normal semigroup is kernel normal (§36). For C_3, we find that A_3^c is not kernel normal while all other normal semigroups (those listed in Figure 36.2) are kernel normal. For C_4, we have Proposition 39.6 below. For background, however, first recall Theorem 26.1, which tells us that the charts in C_4 that are restrictions of permutations in the Klein 4-group V form the 61-semigroup T_{61}. That is,

$$V = \{ \ (1)(2)(3)(4), \ (14)(23), \ (13)(24), \ (12)(34) \ \}$$
$$T_{61} = \{ \ \alpha \in C_4 \mid \alpha \text{ extends to some } \beta \in V \ \}$$

Second, recalling the discussion of X_4 in §29, we see that

$$X_4 = T_{61} \cap I_2.$$

In short, at rank 4, T_{61} is V; at rank 2, it is the "X" in X_4; and at rank ≤ 1, it is I_1 (Figure 39.5):

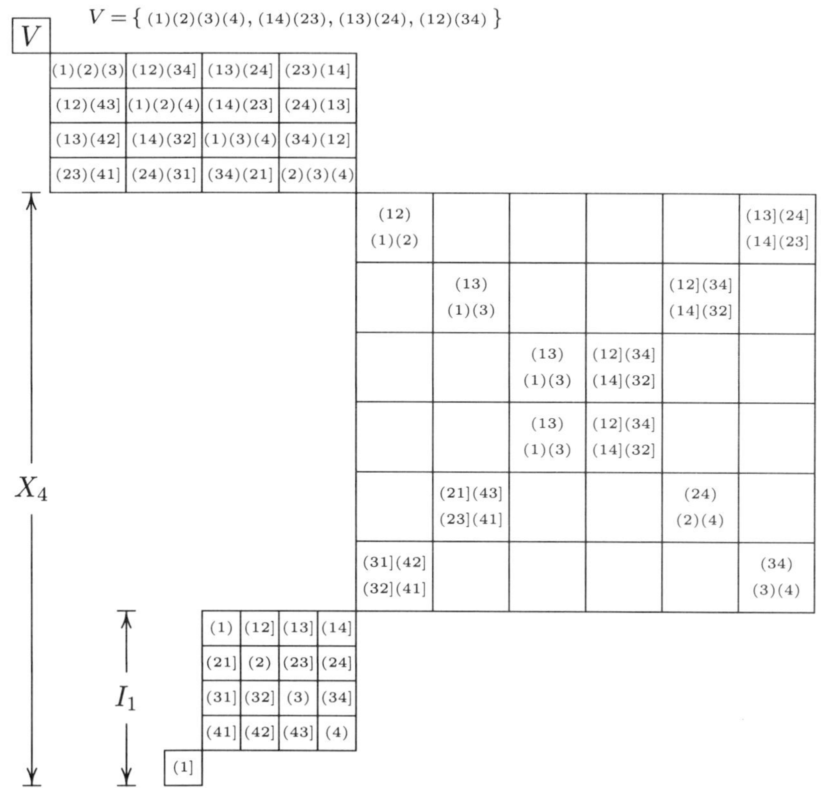

Figure 39.5. The 61-semigroup T_{61} pictured in the egg-box of C_4.

The next proposition follows from 39.1 and 34.2.

39.6 Proposition

Let X_4 denote the X-semigroup on 4 symbols, let I_{r-1} be the ideal of charts of rank $\leq r-1$, let E_4 be the semilattice of idempotents of C_4, and let V be the Klein 4-group. Then K is a normal subsemigroup of C_4 if and only if K is one of the following semigroups:

(1) $I_{r-1} \cup U \cup E_4$, *where* $U = \vee_{1 \leq i \leq \binom{n}{r}} G_i$ *is a disjoint union of conjugate groups defined in* Proposition 32.6,
(2) A_4^c, *the alternating semigroup,*
(3) $E_4 \cup X_4$,
(4) $V \cup (A_4^c - A_4)$,
(5) T_{61}, *the 61-semigroup,*
(6) $T_{57}^1 = (T_{61} - V)^1$, *the 57-semigroup with 1 adjoined,*
(7) $T_{105}^1 = (T_{57} \cup I_2)^1$, *the 105-semigroup with 1 adjoined, and*
(8) $V \cup T_{105}$.

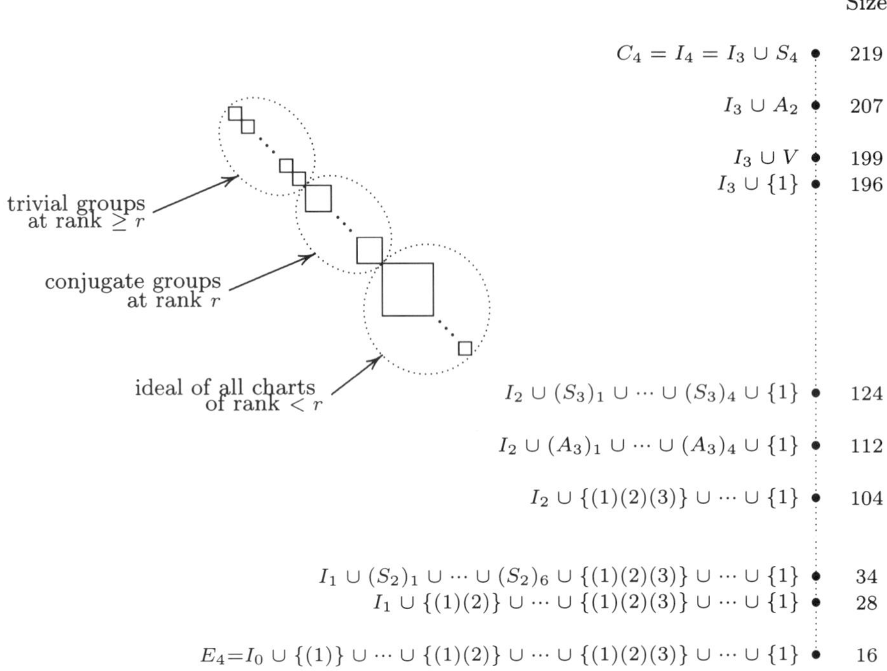

Figure 39.7. A typical egg-box; and the kernel-normal semigroups in C_4.

The "typical egg-box" in Figure 39.7 "pictures" "K" in the following corollary. (To prove 39.8, list the kernel-normal semigroups in 39.6.)

39.8 Corollary

The last seven semigroups listed in 39.6 *are the only normal semigroups in C_4 that are not kernels of any congruences on C_4. Moreover, a normal semigroup K of C_4 is the kernel of a congruence on C_4 if and only if there exists an r, $0 \leq r \leq 4$, such that $K = I_{r-1} \cup U \cup E_4$, where $U = \vee\{\, G_i \mid 1 \leq i \leq \binom{4}{r} \,\}$ is a disjoint union of conjugate groups defined in* Proposition 32.6.

The chain of kernel-normal semigroups in C_4 is pictured in Figure 39.7, where, for example, $(S_2)_1 \cup \cdots \cup (S_2)_6$ denotes the disjoint union $\vee\{\, G_i \mid 1 \leq i \leq \binom{4}{2} \,\}$ of conjugate groups G_i, each isomorphic to S_2.

§40 Counting Congruences

It follows from 38.3 and 39.2 that a congruence ρ on C_n is determined by two parameters, namely, a rank $r \geq 1$, which specifies the $\mathcal{D}$ class D_r

containing the disjoint union $\vee\{\, G_i \mid 1 \le i \le \binom{n}{r} \,\}$, and a group G, which specifies the isomorphism class of the conjugate groups G_i. To make use of this observation, we shall use "ρ^{G_r}" to denote the congruence corresponding to the kernel normal semigroup $K = I_{r-1} \cup \vee\{\, G_i \mid 1 \le i \le \binom{n}{r} \,\} \cup E_n$. As to the kinds of groups that we have for G_r, we see that G_r could be the trivial group 1_r, the symmetric group S_r, the alternating group A_r, or, in the special case when $r = 4$, the Klein 4-group V. With these G_r groups, and with ι_n denoting the identity congruence and $\omega_n = C_n \times C_n$ the universal congruence, the inclusion chain of congruences on C_n may be expressed as follows: For $n \ge 5$,

$$\begin{aligned}
\iota_n = \rho^{S_1} \subset \rho^{1_2} \subset \rho^{S_2} \subset \rho^{1_3} \subset \rho^{A_3} \subset \rho^{S_3} \subset \rho^{1_4} \subset \rho^{V} \subset \rho^{A_4} \subset \rho^{S_4} \\
\subset \rho^{1_5} \subset \rho^{A_5} \subset \rho^{S_5} \subset \cdots \\
\subset \rho^{1_{n-1}} \subset \rho^{A_{n-1}} \subset \rho^{S_{n-1}} \\
\subset \rho^{1_n} \subset \rho^{A_n} \subset \rho^{S_n} \subset \omega_n.
\end{aligned}$$

And for the first four cases ($n < 5$), taken in order:

$$\begin{aligned}
&\iota_1 \subset \omega_1 \\
&\iota_2 = \rho^{S_1} \subset \rho^{1_2} \subset \omega_2 \\
&\iota_3 = \rho^{S_1} \subset \rho^{1_2} \subset \rho^{S_2} \subset \rho^{1_3} \subset \rho^{A_3} \subset \rho^{S_3} \subset \omega_3 \\
&\iota_4 = \rho^{S_1} \subset \rho^{1_2} \subset \rho^{S_2} \subset \rho^{1_3} \subset \rho^{A_3} \subset \rho^{S_3} \subset \rho^{1_4} \subset \rho^{V} \subset \rho^{A_4} \subset \rho^{S_4} \subset \omega_4
\end{aligned}$$

From these inclusion strings, we may easily count congruences — C_1 has two, C_2 has four, C_3 has seven, and, for $n \ge 4$, C_n has $3n - 1$.

§41 Comments

For an arbitrary inverse semigroup S, there are two standard abstract characterizations of its congruences. One is given in terms of a *congruence pair* (K, τ), where K is a normal subsemigroup of S, where τ is a normal congruence on the semilattice of idempotents E_S of S, and where K and τ satisfy certain "intertwining" conditions. The other is given in terms of a *kernel normal system*, which is a pairwise disjoint collection of inverse semigroups satisfying certain conditions. The approach taken in this chapter must obviously involve aspects of both: If a normal semigroup K in C_n is kernel normal with partition $K = I \cup G_1 \cup \cdots \cup G_t$, then the τ in the corresponding congruence pair (K, τ) for C_n is the Rees congruence as determined in 37.3. Furthermore, the partition $K = I \cup G_1 \cup \cdots \cup G_t$ may be viewed as a kernel normal system $\{I, G_1, \ldots, G_t\}$. A systematic exposition of kernel normal systems and congruence pairs may be found in Chapter III of Petrich [1]. For congruence pairs in the wider class of regular semigroups, see Pastijn and Petrich [1].

Congruences on semilattices have been studied by Papert [1], Freese and Nation [1], and Hall [1]. In this chapter, Proposition 37.2 is standard and Proposition 37.3 is due to the author.

Congruences on symmetric inverse semigroups were first studied in 1953 by Liber [1], who used techniques very similar to those used in 1952 by Malcev [1] for characterizing the congruences on full transformation semigroups. An updated approach, with some new insights, was provided in 1973 by Scheiblich [1], who used kernel normal systems and Brandt semigroups to obtain our (§40) chain of congruences for C_n. An approach parallel to Scheiblich's was used by the author [3] to determine the chain of congruences for the alternating semigroups A_n^c.

CHAPTER 9

Presentations of Symmetric Inverse Semigroups

In 1895, E. H. Moore [1] formulated a presentation of S_n that has $n-1$ generators $t_1, \ldots, t_{n-1}$ and defining relations

$$t_i^2 = (t_i t_{i+1})^3 = (t_i t_j)^2 = 1, \quad \text{where } 1 \leq i < n \text{ and } |i-j| > 1.$$

In this chapter, we "extend" Moore's presentation to one for C_n: By augmenting $t_1, \ldots, t_{n-1}$ with one additional generator e, and by increasing the list of $\{t_i\}$-relations with

$$e^2 = e, \quad (et_1)^2 = (et_1)^3, \quad \text{and} \quad et_i = t_i e, \text{ for } 2 \leq i < n,$$

we obtain an inverse monoid presentation of C_n.

That C_n has such a set $X = \{t_1, \ldots, t_{n-1}, e\}$ of generators is easily demonstrated (paragraph containing (42.10)). That C_n has such a presentation is not so easily demonstrated: Following a brief review of the basics in §42, we define, 42.5 and 42.8, respectively, a free inverse monoid M_X and the "representing quotient" M_X/ρ. We then study M_X/ρ in §43; and define the isomorphism $\psi : M_X/\rho \to C_n$ in §44. The main result is Theorem 44.3.

Our approach for C_n runs parallel to the one that we shall use in Chapter 10, where we similarly "extend" E. H. Moore's presentation of the alternating group A_n to one for the alternating semigroup A_n^c.

§42 Review: Free Algebras, Presentations, Constructions

In addition to binary operations $S \times S \to S$, we frequently encounter *unary operations* $S \to S$. For example, we have the unary bijection $^{-1}$: $G \to G$ that takes each element g of a group G to its inverse g^{-1}. That this operation satisfies $(ab)^{-1} = b^{-1}a^{-1}$ and $(g^{-1})^{-1} = g$ motivates the idea of an involution: If S is a semigroup, then a bijection $^{-1} : S \to S$ is an *involution* if $^{-1}$ is an *anti-homomorphism*

$$(ab)^{-1} = b^{-1}a^{-1} \quad (a, b \in S)$$

whose square $(^{-1})^2$ is the identity map $1_S : S \to S$. More generally, if $n \in \{1, 2, \ldots\}$ and if S^n denotes the n-fold cartesian product $S \times \cdots \times S$, then we say that an *operation* $S^n \to S$ has *arity* n; and we call a set S together with possibly several operations of various arities an *algebra*. A *class* $\mathcal{A}$ *of algebras* is one where all of its objects are of the same kind. A *homomorphism* $S \to T$ of such objects is a map that respects all operations.

42.1 Definition

For a class $\mathcal{A}$ of algebras and a nonempty set X, consider a map $\iota : X \to F \in \mathcal{A}$. Then the pair (F, ι) is a free $\mathcal{A}$-algebra on X if for each $A \in \mathcal{A}$ and each $\phi : X \to A$, there exists a unique homomorphism $\psi : F \to A$ such that $\iota \circ \psi = \phi$:

$$\begin{array}{ccc} & & F \\ & \overset{\iota}{\nearrow} & \downarrow \psi \\ X & \xrightarrow[\phi]{} & A \end{array} \tag{42.2}$$

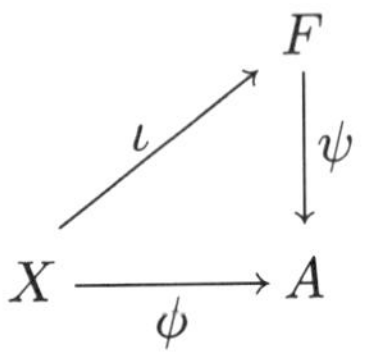

We say that F is a free $\mathcal{A}$-algebra when an ι exists such that (F, ι) is a free $\mathcal{A}$-algebra on some X. The set X is usually taken as a subset of F and is sometimes called a basis for F.

Congruences and Presentations

Recall that a *congruence* ρ on a semigroup S is an equivalence relation $\rho \subset S \times S$ such that

$$(a, b), (a', b') \in \rho \text{ implies } (aa', bb') \in \rho,$$

and that for each congruence ρ on S, the set S/ρ of equivalence classes becomes a semigroup with multiplication

$$[a][b] = [ab] \quad \text{or} \quad (a\rho)(b\rho) = (ab)\rho,$$

which is well-defined because ρ is a congruence. Recall also that there are two constructions of congruences: One is induced from a homomorphism $\psi : S \to T$ between semigroups — define $(a, b) \in \rho$ when $a\psi = b\psi$. This $\rho = \psi \circ \psi^{-1}$ is called the $\ker \psi$, and $S/\ker \psi$ is isomorphic to $S\psi$. The second construction of congruences on a semigroup S is induced from an arbitrary relation $R \subset S \times S$ — since the intersection of congruences is a congruence, the *congruence* ρ *generated by a subset* R *of* $S \times S$ is $\rho = \cap\{\sigma \subset S \times S \mid \sigma$ is a congruence and $R \subset \sigma\}$.

For $R \subset S \times S$ and the congruence ρ generated by R, if $c, d \in S$ satisfy both $c = xpy$ and $d = xqy$ for some $x, y \in S^1$, where either $(p, q) \in R$ or $(q, p) \in R$, then we shall write "$c \to d$" and say that c is connected to d

by an *elementary R-transition*. That these R-transitions may be used to characterize the congruence ρ generated by R is formalized in the following proposition, whose proof is standard and therefore omitted.

42.3 Proposition

Let R be a relation on a semigroup S, and let ρ be the congruence on S generated by R. For any $a, b \in S$, we have $a\rho = b\rho$ if and only if either $a = b$ or there exists a $k \in \{1, 2, \dots\}$ and a sequence

$$a = w_1 \to w_2 \to \cdots \to w_k = b$$

of elementary R-transitions connecting a to b.

In particular, for an index set I and a free semigroup F on X, let $a_i, b_i \in F$ $(i \in I)$ and suppose that $\Delta = \{\, a_i = b_i \,\}_{i \in I}$ is a family of equations. Define ρ to be the congruence generated by $R = R_\Delta = \{(a_i, b_i) \,|\, i \in I\} \subset F \times F$. Then a semigroup S is defined (or *presented*) by the set X of *generators* and the set $\Delta = \{\, a_i = b_i \,|i \in I\}$ of *relations* if it is isomorphic to the factor semigroup F/ρ. In this case we say that S has the *presentation* $\langle\, X \,|\, a_i = b_i; i \in I \,\rangle = \langle\, X \,|\, \Delta \,\rangle$. Moreover, since Δ determines the relation R_Δ, we often define ρ by simply saying *let ρ be the congruence generated by the relations $a_i = b_i$ $(i \in I)$*.

A similar development holds for groups. A group G is defined (or *presented*) by a set X of *generators* and *relations* $\Delta = \{\, r_j = 1 \,|\, j \in J\}$ if G is isomorphic to F/H where F is a free group on X and H is the normal subgroup of F generated by $\{\, r_j \,|\, j \in J \,\}$. In this case, we say that G has *presentation* $\langle\, X \,|\, r_j = 1; j \in J \,\rangle$.

Constructions of Free $\mathcal{A}$-Algebras

To construct a free semigroup on a nonempty set X, we call X an *alphabet* and then refer to its members as *letters*. A *word* $w = x_1 \cdots x_n$ over X is any finite (possibly empty) sequence of letters of X. The *empty word* "$\emptyset$" is the empty set (the word with no letters). Letting X^* denote the set of all words over X, we define multiplication $X^* \times X^* \to X^*$ as concatenation:

$$\begin{aligned} (x_1 \cdots x_n)(y_1 \cdots y_m) &= x_1 \cdots x_n y_1 \cdots y_m \\ w\emptyset = \emptyset w &= w \qquad (w \in X^*). \end{aligned}$$

With this multiplication, X^* is a monoid and $X^+ = X^* - \{\emptyset\}$ is a semigroup. One-letter words are identified with letters.

42.4 Proposition

For a nonempty set X, the monoid X^* over X together with the inclusion map $\iota : X \to X^*$ is a free monoid. Moreover, the semigroup X^+ of nonempty words over X together with the inclusion map $X \to X^+$ is a free semigroup.

Proof. Let S be any semigroup and suppose that $\phi : X \to S$ is any map. Then for $\psi : X^+ \to S$ given by $(x_1x_2\cdots x_k)\psi = x_1\phi x_2\phi\cdots x_k\phi$, we may easily deduce that $\iota \circ \psi = \phi$. In addition, since each word in $F = X^+$ is a unique product of letters in X, the map ψ is well-defined; and since multiplication in F is concatenation, ψ preserves multiplication. To see that ψ is the only morphism that makes the diagram (42.2) commute, note that since ι is inclusion, any morphism ψ' satisfying $\iota \circ \psi' = \phi$ must agree with ψ on the generators (and therefore all) of F. This finishes the proof that (X^+, ι) is a free semigroup with basis X. The proof in the monoid case is similar, we merely need to map the empty word via the formula $\emptyset\psi = 1 \in S$. □

For free semigroups with involution, let $(S, \cdot, ^{-1})$ denote a semigroup S with multiplication "$\cdot$" and involution "$()^{-1}$". If S has an identity, then $(S, \cdot, ^{-1})$ is a *monoid with involution*. For example, an inverse monoid M is a monoid with involution since $a \mapsto a^{-1}$ $(a \in M)$ is an involution.

42.5 Definition (M_X)

Let X and X' be nonempty disjoint sets, and let $\theta : X \to X'$ be a bijection. Define $Y = X \cup X'$, and let Y^* denote the set of all words (including the empty word) with letters in Y under concatenation. Define a unary operation $^{-1} : Y^* \to Y^*$ by

$$x^{-1} = \begin{cases} x\theta & \text{if } x \in X \\ x\theta^{-1} & \text{if } x \in X' \end{cases} \qquad \emptyset^{-1} = \emptyset,$$

and

$$(x_1x_2\cdots x_n)^{-1} = x_n^{-1}x_{n-1}^{-1}\cdots x_1^{-1}.$$

Then the operation "$()^{-1}$" is an involution, and $Z = (Y^*, \cdot, ^{-1})$ is the monoid Y^* with involution $()^{-1}$. For this monoid Z, the Wagner Congruence τ is the congruence generated by the relation

$$\{\, (uu^{-1}u, u) \mid u \in Z \,\} \;\cup\; \{\, (uu^{-1}vv^{-1}, vv^{-1}uu^{-1}) \mid u, v \in Z \,\}.$$

The quotient monoid Z/τ is denoted M_X.

42.6 Proposition

The pair $(Z, 1_X)$, where $1_X : X \to Z$ identifies each letter x with the word $x \in Y^*$, is a free monoid with involution on X; and the pair $(M_X, \tau^\natural|_X)$, where $\tau^\natural|_X : X \to M_X$ maps each $x \in X$ to the class $x\tau$ containing the word $x \in Y^*$, is a free inverse monoid on X.

Proof. Let M be any monoid with involution, which we shall also denote as $()^{-1}$; and suppose that $\phi : X \to M$ is any mapping. First, on $Y = X \cup X'$ define

$$y\psi = \begin{cases} x\phi & \text{if } y = x \in X, \\ (x\phi)^{-1} & \text{if } y = x\theta \in X'. \end{cases}$$

Then on Y^*, define $\emptyset\psi = 1 \in M$ and $(y_1 \cdots y_k)\psi = (y_1\psi)\cdots(y_k\psi)$. For this $\psi : Z \to M$, we clearly have $1_X \circ \psi = \phi$. In addition, since each word in the underlying set Y^* of Z is a unique product of letters in Y, the map ψ is well-defined; and since multiplication in Z is concatenation, ψ preserves multiplication. That ψ also respects involution is a result of $(y^{-1})\psi = (y\psi)^{-1}$ for each $y \in Y$, which follows from the definition of ψ and the meaning of y^{-1} in Z. That ψ is the only morphism such that $1_X \circ \psi = \phi$, notice that any morphism $\psi' : Z \to M$ such that $1_X \circ \psi' = \phi$ agrees with ψ on $X \cup X'$ (and therefore all of Z).

Turning to $M_X = Z/\tau$, we shall use "$\equiv$" to mean equality modulo τ. We first show that M_X is an inverse monoid by showing that it is regular and that its idempotents commute. From the definition of τ, we see that M_X is regular because $w\tau \in Z/\tau$ implies $w^{-1}ww^{-1} \equiv w^{-1}$. And for an idempotent $e\tau$, if $u = e \in Y^*$ and if $v = e^{-1} \in Y^*$, then $uu^{-1}vv^{-1} \equiv vv^{-1}uu^{-1}$ implies $ee^{-1}e^{-1}e \equiv e^{-1}eee^{-1}$, showing that

$$\begin{aligned} e \equiv e^2 \equiv (ee^{-1}e)(ee^{-1}e) &\equiv e(e^{-1}eee^{-1})e \\ &\equiv e(ee^{-1}e^{-1}e)e \equiv (ee^{-1})(e^{-1}e) \equiv ww^{-1}. \end{aligned}$$

Therefore $e \equiv ww^{-1}$ for some $w \in Y^*$ and we conclude, from the definition of τ, that idempotents in Z/τ commute. So $Z/\tau = M_X$ is an inverse semigroup with identity $\emptyset\tau = \emptyset$, i.e., it is an inverse monoid.

To see that M_X is also a free inverse monoid, let M be any inverse monoid and suppose that $\phi : X\tau^\natural \to M$ is any mapping. (Since X generates Z, it is clear that $X\tau^\natural$ generates $Z/\tau = M_X$.) We need to define a $\psi : Z/\tau \to M$ for which the following diagram commutes:

(42.7)

$$\begin{array}{ccccc} & & & & Z/\tau \\ & & & \nearrow & \downarrow \psi \\ X & \xrightarrow[\tau^\natural|_X]{} & X\tau^\natural & \xrightarrow[\phi]{} & M. \end{array}$$

Since $\tau^\natural|_X \circ \phi : X \to M$ is a map where M is a monoid with involution, and since $(Z, 1_X)$ is a free monoid with involution, there is a unique morphism $\phi_1 : Z \to M$ that respects involution and makes the following diagram commute:

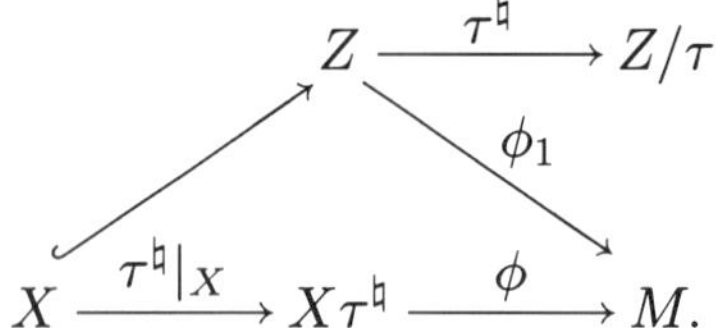

Using ϕ_1, we define $(w\tau)\psi = w\phi_1$ for each $w\tau \in Z/\tau$, obtaining the $\psi : Z/\tau \to M$ required in diagram (42.7). Now ψ is well-defined since $\tau \subset \phi_1 \circ \phi_1^{-1}$. That ψ respects both multiplication and the involution of taking inverses follows from the fact that ϕ_1 respects multiplication and involution. In other words, ψ is an inverse monoid morphism. Considering that $X\tau^\natural$ generates Z/τ, we see that ψ is the only such morphism that makes (42.7) commute. □

42.8 Definition (M_X/ρ)

For $X = \{t_1, t_2, \ldots, t_{n-1}, e\}$, let M_X denote the free inverse monoid defined in 42.5. *Let ρ be the congruence on M_X generated by the relations*

$$t_i^2 = (t_i t_{i+1})^3 = (t_i t_j)^2 = 1, \text{ where } 1 \le i < n \text{ and } |i-j| > 1,$$
$$e^2 = e, \quad (et_1)^2 = (et_1)^3, \quad \text{and} \quad et_i = t_i e \text{ when } i > 1.$$

Then M_X/ρ will denote the quotient semigroup whose members are ρ-classes of elements in M_X.

To construct a free group, we again start with the $Z = (Y^*, \cdot, ^{-1})$ as defined in 42.5, but in this case we use "1" to denote the empty word $\emptyset$, and we also use the congruence σ generated by the relation

$$R = \{\, (uu^{-1}, 1) \mid u \in Z \,\} \ \cup \ \{\, (u^{-1}u, 1) \mid u \in Z \,\}.$$

Then each equivalence class $w\sigma$ contains a unique "reduced word" $r(w)$, and it is these words that comprise the free group G_X with basis X.

In detail, letting $w = y_1 y_2 \cdots y_k \in Y^*$ be a word of *length* $|w| = k$, we define a *reduction*

$$w \mapsto y_1 y_2 \cdots y_{i-1} y_{i+2} \cdots y_k$$

when $y_{i+1} = y_i^{-1}$. (Note that a reduction is, in effect, an elementary R-transition.) A word w that has no reduction is a *reduced word*. Clearly, any word $w \in Y^*$ is either a reduced word, in which case we write $r(w) = w$, or there exists a finite sequence $w \mapsto w_1 \mapsto \cdots \mapsto w_k$ of reductions where $r(w) = w_k$ is a reduced word. That $r : Y^* \to Y^*$ is a function may be deduced by showing that for distinct reduced words $u \neq v$, the existence of a finite sequence $u = w_1 \mapsto w_2 \mapsto \cdots \mapsto w_n = v$ of reductions having minimal $\sum |w_i|$ is impossible. It follows that $r(w)$ is independent of the order in which the reductions are performed.

We define $G_X = \{\, w \in Y^* \mid r(w) = w\}$ and a multiplication on G_X via

$$w \cdot u = r(wu) \quad (w, u \in G_X).$$

Since concatenation of two reduced words may not be a reduced word, verifying associativity is somewhat case intensive. Nevertheless, the product is associative, $r(1) = 1 \in G_X$, and each $w \in G_X$ has an inverse $w^{-1} \in G_X$, making G_X a group.

42.9 Proposition

The group G_X together with the inclusion map $X \to G_X$ is a free group.

When speaking of the elements of the free group G_X on X, the words in Y^* (even those that are not reduced) serve as names for the elements of G_X. So for any word $w \in Y^*$, "the group element w" is well defined.

Group Presentations extended to Inverse Monoid Presentation

Recall that we are concerned with extending a group presentation of S_n to an inverse monoid presentation of C_n. Beginning with S_n, we see that the $n-1$ transpositions $t_1 = (1,2)$, $t_2 = (2,3)$, $\ldots$, $t_{n-1} = (n-1, n) \in S_n$ may be used to verify that S_n has a set $X = \{t_1, \ldots, t_{n-1}\}$ of generators that satisfy the relations

$$t_i^2 = (t_i t_{i+1})^3 = (t_i t_j)^2 = 1, \qquad \text{where } 1 \leq i < n \text{ and } |i-j| > 1.$$

In the C_n case, we may augment these transpositions $t_1, \ldots, t_{n-1}$ with the rank $n-1$ idempotent $e = (1](2)\cdots(n)$ to verify that C_n has a set $X \cup \{e\}$ of generators that satisfy the relations

$$\begin{aligned} t_i^2 = (t_i t_{i+1})^3 = (t_i t_j)^2 = 1, &\qquad \text{where } 1 \leq i < n \text{ and } |i-j| > 1, \\ e^2 = e, \quad (et_1)^2 = (et_1)^3, &\quad \text{ and } \quad et_i = t_i e, \text{ for } 2 \leq i < n. \end{aligned} \tag{42.10}$$

In effect, S_n has a group presentation $\langle X \mid \Delta \rangle$, and we shall show that C_n has the inverse monoid presentation $\langle X \cup \{e\} \mid \Delta \cup \Delta_1 \rangle$ where $\Delta \cup \Delta_1$ is the set of relations listed in (42.10). We notice, for each relation $u = v$ in Δ_1, that both u and v contain the letter e. This observation shows that the extension of the presentation $\langle X \mid \Delta \rangle$ of S_n to $\langle X \cup X_1 \mid \Delta \cup \Delta_1 \rangle$ of C_n is a special case of the following.

42.11 Proposition

Let $G = (G, \cdot, ^{-1})$ be an inverse monoid with presentation $\langle X \mid \Delta \rangle$, and let $M = (M, \cdot, ^{-1})$ be an inverse monoid with presentation $\langle X \cup X_1 \mid \Delta \cup \Delta_1 \rangle$. If X and X_1 are disjoint, and if for each relation $u = v$ in Δ_1, both u and v contain a letter from $X_1 \cup X_1^{-1}$, then there exists a monomorphism $G \to M$ that fixes each generator $x \in X$.

Proof. Let σ be the congruence on the free inverse monoid M_X induced by Δ, let σ_1 be the congruence on the free inverse monoid $M_{X \cup X_1}$ induced by Δ_1, and suppose that ρ is the congruence on $M_{X \cup X_1}$ induced by $\Delta \cup \Delta_1$. Then $\sigma_1 \subset \rho$ provides a well defined map $\alpha : M_{X \cup X_1}/\sigma_1 \to M_{X \cup X_1}/\rho$ given by $(w\sigma_1)\alpha = w\rho$. In addition, since both σ_1 and ρ are congruences, it is easy to show that α is a homomorphism. Then for $Y = X \cup X'$ as in 42.5, we may also show that α is one-one on $K = \{ w\sigma_1 \mid$ every letter in w is a letter in Y $\}$. (If $w\rho = (w\sigma_1)\alpha = (u\sigma_1)\alpha = u\rho$, where all letters in both w and u are letters in Y, then, by the property of Δ_1, any sequence $w = w_1 \to \cdots \to w_k = u$ of elementary $(\Delta \cup \Delta_1)$-transitions is necessarily a sequence of Δ-transitions.) It follows that $\alpha|_K$ imbeds K into $M_{X \cup X_1}/\rho$ while fixing each generator $x \in X$. And since $\beta : M_X/\sigma \to K$ given by $w\sigma = w\sigma_1$ is an isomorphism that also fixes each generator $x \in X$, we have that $\beta \circ \alpha$ imbeds M_X/σ into $M_{X \cup X_1}/\rho$ while fixing each generator $x \in X$. The desired result now follows since G and M are isomorphic to M_X/σ and $M_{X \cup X_1}/\rho$, respectively. □

§43 Technical Lemma

Because each generator $x \in X$ has been identified with its corresponding ρ-class $x\rho \in M_X/\rho$, we may consider any string $x_1 \cdots x_k$ of generators as a representative of some ρ-class. In particular, $t_i \in X$ and $e \in X$ represent the ρ-classes $[t_i]$ and $[e]$, respectively. And to show that C_n has the presentation $\langle \{ t_1, \ldots, t_{n-1}, e\} \mid \Delta' \rangle$, where Δ' consists of the relations (42.10), we shall construct an isomorphism $\psi : M_X/\rho \to C_n$ that extends $[t_i] \mapsto (i, i+1) \in C_n$ and $[e] \mapsto (1](2) \cdots (n)$. But first, we consider representatives of ρ-classes that will correspond to rank $n-1$ idempotents in C_n: We define e_i, $1 \leq i \leq n$, such that $[e_i]$ will correspond to $(1) \cdots (i-1)(i](i+1) \cdots (n)$,

$$(43.1) \qquad \begin{aligned} e_1 &= e \\ e_2 &= t_1 e_1 t_1 \\ &\vdots \\ e_n &= t_{n-1} e_{n-1} t_{n-1}. \end{aligned}$$

Intuitively speaking, these forms were a result of the realization that all rank $n-1$ idempotents have the same path structure — each is conjugate to $(1](2)\cdots(n)$. For instance, $e_2 = t_1e_1t_1$ corresponds to the conjugate $(1)(2](3)\cdots(n)$ of $(1](2)\cdots(n)$, and is motivated by

$$(1)(2](3)\cdots(n) = (12)\circ(1](2)\cdots(n)\circ(12).$$

Other ρ-equivalent words that we shall need are itemized in Lemma 43.2 below. These equivalencies may be understood (on first pass) by placing them in the C_n context. For instance, to take one of many examples, in trying to understand part (5) of 43.2, keep in mind that t_{ik} is just a "word" that corresponds to the transposition (i,k).

43.2 Lemma

Let M_X be the free inverse monoid generated by $X = \{t_1, \dots, t_{n-1}, e\}$, and let ρ be the congruence on M_X generated by the relations given in (42.10). Then when "equality" is understood to mean "equality modulo ρ," we have the following:

(1) *Each $e_i\rho$, $1 \le i \le n$, is an idempotent in M_X/ρ.*
(2) *In addition to the relations in (42.10), we have relations $(et_1)^2 = (t_1e)^2 = (t_1e)^3$.*
(3) *For relations among the e_i and the t_i we have*

$$\begin{aligned} t_ie_i &= e_{i+1}t_i && \textit{when } 1 \le i < n, \\ e_it_i &= t_ie_{i+1} && \textit{when } 1 \le i < n, \\ e_it_{i+1} &= t_{i+1}e_i && \textit{when } 1 \le i < n, \quad \textit{and} \\ t_je_i &= e_it_j && \textit{when } |i-j| > 1. \end{aligned}$$

(4) *For any word w in M_X, we have $w = t_{j_1}t_{j_2}\cdots t_{j_q}e_{i_1}e_{i_2}\cdots e_{i_k}$.*
(5) *For $1 \le i < n$, let both $t_{i,i+1}$ and $t_{i+1,i}$ denote t_i. And for $k > i+1$, let*

$$\begin{aligned} &t_{ik} \ \textit{denote} \ t_it_{i+1}\cdots t_{k-2}t_{k-1}t_{k-2}\cdots t_{i+1}t_i \ \textit{and let} \\ &t_{ki} \ \textit{denote} \ t_{k-1}t_{k-2}\cdots t_{i+1}t_it_{i+1}\cdots t_{k-2}t_{k-1}. \end{aligned}$$

Then for any i,k with $i \ne k$, we may conclude that $t_{ik} = t_{ki}$ and that $e_ie_kt_{ik} = t_{ik}e_ie_k = e_ie_k$.

Proof. (Of (1)): We use induction on i, $1 \le i \le n$. Clearly, $e_1\rho = e\rho$ is an idempotent since $(e^2, e) \in \rho$. For $i > 1$, consider

$$e_ie_i = (t_{i-1}e_{i-1}t_{i-1})(t_{i-1}e_{i-1}t_{i-1}) = t_{i-1}e_{i-1}^2t_{i-1}.$$

Thus, if $e_{i-1}\rho$ is an idempotent, then so is $e_i\rho$. (Of (2)): The first equality holds since idempotents in M_X/ρ commute, i.e.,

$$t_1et_1e = e_2e = ee_2 = et_1et_1.$$

The second equality holds since M_X is an inverse monoid and hence M_X/ρ is also. That is, we are given that $(et_1)^2 = (et_1)^3$, and so the inverse $(t_1e)^2$ of $(et_1)^2$ must be ρ-equivalent to the inverse $(t_1e)^3$ of $(et_1)^3$. (Of (3)): The validity of the first set of relations follows from $t_ie_i = t_ie_i(t_it_i) = e_{i+1}t_i$. Similarly, the second set is valid. For the third set, consider the following expansion of e_i,

$$e_i = t_{i-1}\cdots t_1et_1\cdots t_{i-1}.$$

Since t_{i+1} commutes with each letter in this expansion of e_i, it also commutes with e_i. For the fourth set, first assume that $j > i+1$, and then replace t_{i+1} in the previous sentence with t_j. Next, consider the other case, i.e., $i > j+1$. Since t_j commutes with t_k when $k > j+1$,

$$\begin{aligned} t_je_i &= t_j(t_{i-1}\cdots t_1et_1\cdots t_{i-1}) \\ &= t_{i-1}\cdots t_{j+2}t_j(t_{j+1}\cdots t_1et_1\cdots t_{j+1})t_{j+2}\cdots t_{i-1} \\ &= t_{i-1}\cdots t_{j+2}t_j(e_{j+2})t_{j+2}\cdots t_{i-1}. \end{aligned}$$

From this, it clearly suffices to consider only the case $i = j+2$: We have

$$\begin{aligned} t_je_{j+2} &= (t_jt_{j+1}t_j)e_jt_jt_{j+1} \\ &= (t_{j+1}t_jt_{j+1})e_jt_jt_{j+1} && \text{since } (t_{j+1}t_j)^3 = 1 \\ &= t_{j+1}t_j(e_jt_{j+1})t_jt_{j+1} && \text{since } e_jt_{j+1} = t_{j+1}e_j \\ &= t_{j+1}t_je_j(t_jt_{j+1}t_j) && \text{since } (t_{j+1}t_j)^3 = 1 \\ &= e_{j+2}t_j. \end{aligned}$$

(Of (4)): Apply statement (3) (move the e's to the right). (Of (5)): To see that $t_{ik} = t_{ki}$, first use the relations $(t_jt_{j+1})^3 = 1$, $1 \le j < n$, to show that

$$t_jt_{j+1}t_j = t_{j+1}t_jt_{j+1}, \qquad 1 \le j < n. \tag{43.3}$$

By iterating with (43.3), the following models may be used to construct a detailed proof of $t_{ik} = t_{ki}$: $t_{13} = t_1t_2t_1 = t_2t_1t_2 = t_{31}$, and

$$\begin{aligned} t_{14} &= t_1(t_2t_3t_2)t_1 \\ &= t_1(t_3t_2t_3)t_1 \\ &= t_3(t_1t_2t_1)t_3 \\ &= t_3(t_2t_1t_2)t_3 = t_{41}. \end{aligned}$$

Next, by expanding t_{ik} and moving (via statement (3)) e_k and e_i to the right, we see that it is straightforward to show that $e_i e_k t_{ik} = t_{ik} e_i e_k$. To see $e_i e_k t_{ik} = e_i e_k$ is more involved. We shall need statements (43.3), (43.4), and (43.5). Thus, consider

$$(43.4) \qquad t_1 e t_1 e = e t_1 e.$$

To see that (43.4) is true, first apply part (2), i.e., $(t_1 e)^2 = (e t_1)^2$:

$$(43.5) \qquad (t_1 e t_1 e) t_1 = (e t_1 e t_1) t_1 = e t_1 e.$$

Then (43.4) follows by first applying $(t_1 e)^2 = (t_1 e)^3$, and then (43.5) in the calculation

$$t_1 e t_1 e = (t_1 e t_1 e t_1) e = (e t_1 e) e = e t_1 e.$$

Next, we use (43.4) to show that for $k > 1$

$$(43.6) \qquad e_1 t_{1k} e_1 = e_1 e_k.$$

If $k = 2$ in (43.6), then (43.6) is equivalent to (43.4). To see that (43.6) is also true when $k > 2$, use $t_{1k} = t_{k1}$ and calculate that

$$\begin{aligned}
e_1 t_{1k} e_1 &= e_1 t_{k1} e_1 \\
&= e(t_{k-1} \cdots t_2 t_1 t_2 \cdots t_{k-1}) e \\
&= e t_{k-1} \cdots t_2 t_1 e t_2 \cdots t_{k-1} \\
&= t_{k-1} \cdots t_2 (e t_1 e) t_2 \cdots t_{k-1} \\
&= t_{k-1} \cdots t_2 (t_1 e t_1 e) t_2 \cdots t_{k-1} \qquad \text{from (43.4)} \\
&= t_{k-1} \cdots t_2 t_1 e t_1 t_2 \cdots t_{k-1} e \\
&= e_k e_1 = e_1 e_k.
\end{aligned}$$

Having proved statements (43.4), (43.5), and (43.6), we are now ready to show $e_i e_k t_{ik} = e_i e_k$ where $k > i$. For our first case, $k = 2$ and $i = 1$, consider

$$e_i e_k t_{ik} = e_1 e_2 t_1 = e_2 e_1 t_1 = t_1 e t_1 e t_1.$$

And first apply (43.5) to the term on the right, and then apply (43.4). For our second case, $k > 2$ and $i = 1$, use

$$e_i e_k t_{ik} = e_1 e_k t_{1k} = e_1 t_{1k} e_1$$

and apply (43.6) to the term on the right. For our third case, $k = i + 1$ and $i > 1$, we have

$$\begin{aligned}
e_i e_{i+1} t_{i,i+1} &= e_i (e_{i+1} t_i) = e_i t_i e_i \\
&= t_{i-1} \cdots t_1 e (t_1 \cdots t_{i-1} t_i t_{i-1} \cdots t_1) e t_1 \cdots t_{i-1} \\
&= t_{i-1} \cdots t_1 (e t_{1,i+1} e) t_1 \cdots t_{i-1} \\
&= t_{i-1} \cdots t_1 (e_1 e_{i+1}) t_1 \cdots t_{i-1} \qquad \text{from (43.6)} \\
&= (t_{i-1} \cdots t_1 e_1 t_1 \cdots t_{i-1}) e_{i+1} \\
&= e_i e_{i+1}.
\end{aligned}$$

Finally, we assume $k > i+1$ and $i > 1$: We calculate that

$$\begin{aligned}
e_i e_k t_{ik} &= e_i e_k t_{ki} \\
&= e_i(t_{k-1}\cdots t_i t_{i-1}\cdots t_1 e t_1 \cdots t_{i-1} t_i \cdots t_{k-1})(t_{k-1}\cdots t_{i+1} t_i t_{i+1}\cdots t_{k-1}) \\
&= e_i t_{k-1}\cdots t_i t_{i-1}\cdots t_1 e t_1\cdots t_{i-1}(t_i\cdots t_{k-1}t_{k-1}\cdots t_i)t_{i+1}\cdots t_{k-1} \\
&= e_i t_{k-1}\cdots t_i t_{i-1}\cdots t_1 e t_1\cdots t_{i-1}(t_{i+1}\cdots t_{k-1}) \\
&= e_i t_{k-1}\cdots t_i(t_{i+1}\cdots t_{k-1})t_{i-1}\cdots t_1 e t_1\cdots t_{i-1} \qquad \text{since } k-1 \geq i+1 \\
&= e_i t_{ki} e_i \\
&= t_{i-1}\cdots t_1 e(t_1\cdots t_{i-1} t_{ik} t_{i-1}\cdots t_1) e t_1\cdots t_{i-1} \qquad \text{since } t_{ki} = t_{ik} \\
&= t_{i-1}\cdots t_1(e_1 t_{1k} e_1)t_1\cdots t_{i-1} \\
&= t_{i-1}\cdots t_1(e_1 e_k)t_1\cdots t_{i-1} \qquad \text{from (43.6)} \\
&= t_{i-1}\cdots t_1 e_1 t_1\cdots t_{i-1} e_k \qquad \text{since } k-(i-1) > 1 \\
&= e_i e_k.
\end{aligned}$$

This completes all cases where $k > i$. It now follows from $t_{ik} = t_{ki}$ and the commutativity of idempotents in M_X/ρ that $e_i e_k t_{ik} = e_i e_k$ for all $k \neq i$. This finishes the proof. □

§44 Inductive Lemma

For $X = \{t_1, \dots, t_{n-1}, e\}$ and for $\iota : X \to M_X/\rho$ given by $x\iota = [x] \in M_X/\rho$, we have $(M_X/\rho, \iota)$ as the freest inverse monoid that is generated by $X = \{t_1, \dots, t_{n-1}, e\}$ and that satisfies the relations (42.10). We also have $\phi : X \to C_n$ where $t_i\phi = (i, i+1)$, $1 \leq i \leq n-1$, and $e\phi = (1](2)\cdots(n)$. The relative freeness of M_X/ρ, coupled with the fact that $X\phi$ is a generating set of C_n, says that there is a unique epimorphism $\psi : M_X/\rho \to C_n$ such that $(x\iota)\psi = (x\rho)\psi = x\phi$, i.e., such that the following diagram commutes

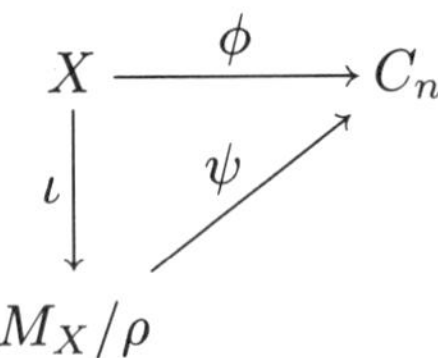

To simplify notation, we shall denote both ι (from X to M_X/ρ) and the natural map $\rho^\natural$ (from M_X to M_X/ρ) as ρ. Moreover, for any $B = \{i_1, i_2, \dots, i_m\} \subset \{1, 2, \dots, n\}$, let

$$e_B = e_{i_1} e_{i_2} \cdots e_{i_m}, \quad e_\emptyset = 1,$$

and then note that e_B is well-defined since idempotents (in M_X/ρ) commute. Also define the composition of $\rho : M_X \to M_X/\rho$ followed by ψ as the "hat" map $\widehat{\ } = \rho \circ \psi$, i.e., $\widehat{w} = (w\rho)\psi$. So when t is a product of elements in $\{t_1, \dots, t_{n-1}\}$, then $\widehat{te_B} = (te_B)\rho\psi = (t)\rho\psi\,(e_B)\rho\psi = \widehat{t}\widehat{e}_B$.

44.1 Lemma

Let $t, t_0 \in \{t_1, \dots, t_{n-1}\}$, and $B \subset N$. If $\widehat{t}\widehat{e}_B = \widehat{t_0}\widehat{e}_B$, then $(te_B, t_0e_B) \in \rho$.

Proof. Let p_m denote the statement "$\widehat{t}\widehat{e}_B = \widehat{t_0}\widehat{e}_B$ when $|B| \leq m \Rightarrow (te_B, t_0e_B) \in \rho$. We show p_0, p_1, and (the implication) "$p_{m-1} \Rightarrow p_m$" are true. If $m = 0$, then $B = \emptyset$ and, consequently, $\widehat{t} = \widehat{t_0}$. It follows, since an application of 42.11 shows that the homomorphism ψ restricted to the subgroup $\langle\{t_i\rho\}\rangle$ of M_X/ρ is an isomorphism onto the subgroup S_n of C_n, that $(t, t_0) \in \rho$. Next, let $m = 1$ and suppose that

$$\widehat{t}\widehat{e}_i = \widehat{t_0}\widehat{e}_i.$$

For the index $i \notin \mathbf{r}(\widehat{t}\widehat{e}_i) = \mathbf{r}(\widehat{t_0}\widehat{e}_i)$, each of the charts $\widehat{t}\widehat{e}_i$ and $\widehat{t_0}\widehat{e}_i$ has rank $n-1$, yielding $\widehat{t} = \widehat{t_0}$. And then the imbedding of $\langle\{t_i\rho\}\rangle$ mentioned above gives $(t, t_0) \in \rho$, which shows that $(te_i, t_0e_i) \in \rho$.

So suppose that $m = |B| \geq 2$, that p_{m-1} is true, and that $\widehat{t}\widehat{e}_B = \widehat{t_0}\widehat{e}_B$. If $x \in \{1, 2, \dots, n\}$ implies $x\widehat{t} = x\widehat{t_0}$, then, by the imbedding of $\langle\{t_i\rho\}\rangle$, we are finished. Otherwise, an $x \notin \mathbf{d}(\widehat{t}\widehat{e}_B) = \mathbf{d}(\widehat{t_0}\widehat{e}_B)$ exists such that

$$k = x\widehat{t} \neq x\widehat{t_0} = i \quad \text{and} \quad i, k \in B. \tag{44.2}$$

Since $|B| \geq 2$, we have the representation $\widehat{t}\widehat{e}_B = \widehat{t}\widehat{e}_i\widehat{e}_k\widehat{e}_L = \widehat{t_0}\widehat{e}_i\widehat{e}_k\widehat{e}_L = \widehat{t_0}\widehat{e}_B$, where $B = L \cup \{i, k\}$, $L \cap \{i, k\} = \emptyset$, and (44.2) holds. From part (5) of 43.2,

$$t_0e_B = t_0e_ie_ke_L = t_0e_ie_kt_{ik}e_L = t_0t_{ik}e_ie_ke_L = t_0t_{ik}e_B.$$

It follows that if $t_* = t_0t_{ik}$, then $t_0e_B = t_*e_B \pmod{\rho}$. So $\widehat{t}$ agrees with $\widehat{t_*}$ on $\mathbf{d}(\widehat{t}\widehat{e}_B) \cup \{x\}$. Thus, if $C = B - \{k\}$, then $\widehat{t}\widehat{e}_C = \widehat{t_*}\widehat{e}_C$, where $|C| = m-1$. Hence, by the inductive hypothesis, $(te_C, t_*e_C) \in \rho$; and so $(te_Ce_k, t_*e_Ce_k) = (te_B, t_*e_B) \in \rho$, showing $t_0e_B = t_*e_B \pmod{\rho}$. □

44.3 Theorem

Let M_X be the free inverse monoid generated by $X = \{t_1, t_2, \dots, t_{n-1}, e\}$, and let ρ be the congruence on M_X generated by the relations

$$t_i^2 = (t_it_{i+1})^3 = (t_it_j)^2 = 1, \quad \textit{where } 1 \leq i < n \textit{ and } |i-j| > 1,$$
$$e^2 = e, \quad (et_1)^2 = (et_1)^3, \quad \textit{and} \quad et_i = t_ie \textit{ when } i > 1.$$

Then M_X/ρ is isomorphic to C_n.

Proof. Let $\psi : M_X/\rho \to C_n$ be the unique morphism defined in the first paragraph of §44. From (4) of 43.2, we may assume that every $w \in M_X$ is in a ρ-class with a word of the form te_B where $e_B = e_{i_1} e_{i_2} \cdots e_{i_m}$ and each e_i is defined via (43.1). Then, the assumption that ψ is not one-one would contradict 44.1. The epimorphism ψ is therefore a isomorphism. □

§45 Comments

The circle of ideas surrounding "generators and defining relations" is both old and new: In modern mathematics, since World War II in particular, *defining relations* are often cast in the context of *solutions to universal mapping problems*. In fact, the commutative diagrams in §42 and §44 are now typical. For a nice overview of these developments, see W. S. Massey's 1967 text [1]. For generators and defining relations in group theory, the 1977 text by R. Lyndon and P. Schupp [1] contains, in addition to several historical references in its preface, many detailed discussions of the kinds of problems encountered in the generator-relations approach to group theory. For progress in the study of free inverse semigroups since the early 1970s, see Chapter 2 of Higgins' text [1]; and for presentations of finite symmetric semigroups, i.e., for the semigroup T_n of all functions from N into N under the usual composition of mappings, see Aĭzenštat [1].

The early work on defining relations for S_n is discussed in Burnside [1], where Moore [1] and Burnside [2] are also referenced. A proof that any set of generators of S_n coupled with any rank $n-1$ member of C_n is a generating set for C_n may be found in Gomes and Howie's 1987 paper [1, Theorem 3.1].

The presentation of C_n presented in this chapter (the relations

$$e^2 = e, \quad (et_1)^2 = (et_1)^3, \quad \text{and} \quad et_i = t_i e \quad (\text{for } 2 \le i < n)$$

augmented with the $\{t_i\}$ relations of S_n are defining relations for C_n) was formulated in 1991 by the author, who later learned that this presentation may be deduced from one found in 1961 by L. M. Popova [1]. Popova's result may be viewed as augmenting the $\{t_i\}$ relations of S_n with the relations

$$e^2 = e, \quad e = t_2 e t_2, \quad e = (t_{n-1} t_{n-2} \cdots t_2) e (t_2 t_3 \cdots t_{n-1}),$$
$$(et_1)^2 = et_1 e, \quad (et_1)^2 = (t_1 e)^2.$$

CHAPTER 10

Presentations of Alternating Semigroups

In 1895, E. H. Moore [1] formulated a presentation of A_n that has $n-2$ generators $s_1, \ldots, s_{n-2}$ and relations

$$s_1^3 = s_i^2 = (s_{i-1}s_i)^3 = (s_j s_k)^2 = 1, \quad \text{where } i > 1 \text{ and } |j-k| > 1.$$

In this chapter, we "extend" Moore's presentation to one for A_n^c: By augmenting $s_1, \ldots, s_{n-2}$ with one additional generator e, and by increasing the list of $\{s_i\}$-relations with

$$e^2 = e, \ (es_1)^3 = (es_1)^4, \ (es_j)^2 = (es_j)^4,$$
$$es_i = s_i s_1^{-1} es_1, \quad \text{where } j > 1 \text{ and } i \geq 1,$$

we obtain an inverse monoid presentation of A_n^c.

That A_n^c has such a set $X = \{s_1, \ldots, s_{n-2}, e\}$ of generators is easily demonstrated (§46). That A_n^c has such a presentation is not so easily demonstrated: Given the free inverse monoid M_X (§42 of the previous chapter), we use the relations given above to generate a congruence σ. Our goal is the construction of an isomorphism $\psi : M_X/\sigma \to A_n^c$. We formally define M_X/σ in §46; we study M_X/σ in §47; and we define the morphism ψ in §48. The main result is Theorem 48.3. (The reader may wish to review §42, where the basic ideas are introduced.)

Our approach for A_n^c runs parallel to the one used in Chapter 9, where E. H. Moore's presentation of S_n was "extended" to one for C_n.

§46 Defining Relations

We may use the $n-2$ permutations $s_1 = (1,2,3)$, $s_2 = (1,2)(3,4)$, $\ldots$, $s_i = (1,2)(i+1, i+2)$, $\ldots$, $s_{n-1} = (1,2)(n-1, n)$ to verify that A_n has a set of generators that satisfy the relations

$$s_1^3 = s_i^2 = (s_{i-1}s_i)^3 = (s_j s_k)^2 = 1, \quad \text{where } i > 1 \text{ and } |j-k| > 1.$$

Turning to the A_n^c case, we begin with the following proposition.

46.1 Proposition

Let $n \geq 3$, let W be any generating set for the alternating group A_n, and let $\alpha \in C_n$ be any even chart of rank $n-1$. Then $X = W \cup \{\alpha\}$ is a generating set for A_n^c.

Proof. Since A_n^c is generated by the totality of rank $n-1$ charts of the form (i,j,k) or $(i,j,k]$, and since W generates $A_n \subset A_n^c$, it suffices to show that every rank $n-1$ chart $(i,j,k]$ is a product of members of $X \cup X^{-1}$. First, since $\alpha \in X$ has rank $n-1$, there is a unique $q \notin \mathbf{d}\alpha$ such that $\alpha\alpha^{-1} = (1)\cdots(q-1)(q](q+1)\cdots(n)$. Second, since $n \geq 3$, we choose $\beta = (kqm) \in A_n$, and then calculate that

$$\varepsilon_k = \beta\alpha\alpha^{-1}\beta^{-1} = (1)\cdots(k-1)(k](k+1)\cdots(n)$$

is a product of members of $X \cup X^{-1}$. Since both (ijk) and β are products of members of W, we calculate that $(ijk] = \varepsilon_k \circ (ijk) = \beta\alpha\alpha^{-1}\beta^{-1} \circ (ijk)$ is a product of members of $X \cup X^{-1}$. □

For $n \geq 3$, it follows from 46.1 that

$$X = \{\, s_1 = (1,2,3), s_2 = (1,2)(3,4), \dots ,\\ s_{n-2} = (1,2)(n-1,n),\ e = (1](2)\cdots(n)\,\}$$

is a generating set for A_n^c. It is readily verified that these generators satisfy the following relations:

$$\begin{aligned} &s_1^3 = s_i^2 = (s_{i-1}s_i)^3 = (s_js_k)^2 = 1, \quad \text{where } i > 1 \text{ and } |j-k| > 1, \\ (46.2) \quad &e^2 = e, \ \ (es_1)^3 = (es_1)^4, \\ &(es_j)^2 = (es_j)^4, \ \ es_i = s_is_1^{-1}es_1, \quad \text{where } j > 1 \text{ and } i \geq 1. \end{aligned}$$

46.3 Definition (M_X/σ)

Let $n \geq 3$. Let M_X be the free inverse monoid generated by $X = \{s_1, s_2, \dots , s_{n-2}, e\}$, and let σ be the congruence on M_X generated by the relations listed in (46.2). Then M_X/σ is the inverse monoid whose members are σ-classes of elements in M_X.

§47 A_n^c Technical Lemma

Each generator $x \in X$ is identified with its corresponding σ-class $x\sigma \in M_X/\sigma$, and we may consider a string $y_1 \cdots y_k$, where $y \in Y^* = X \cup X^{-1}$ of generators as a representative of a σ-class. In particular, $s_i \in X$ and $e \in X$ represent the σ-classes $[s_i]$ and $[e]$, respectively. And to show that A_n^c has the presentation $\langle\{s_1, \dots, s_{n-2}, e\} \mid \Delta'\rangle$, where Δ' consists of the relations (46.2), we shall construct an isomorphism $\psi : M_X/\sigma \to A_n^c$ that extends $[s_i] \mapsto (12)(i+1, i+2) \in A_n^c$ and $[e] \mapsto (1](2)\cdots(n)$.

But first, we consider representatives of σ-classes that will correspond to rank $n-1$ idempotents: We define e_i, $1 \le i \le n$, such that $[e_i]$ will correspond to $(1)\cdots(i-1)(i](i+1)\cdots(n)$,

$$\begin{aligned}
e_1 &= e \\
e_2 &= s_1^{-1} e_1 s_1 \quad \text{(Also, } e_2 = s_2 e_1 s_2 \text{ since, from (46.2), } e s_2 = s_2 e_2) \\
e_3 &= s_1^{-1} e_2 s_1 \\
e_4 &= s_2 e_3 s_2 \\
e_5 &= s_3 e_4 s_3 \\
&\vdots \\
e_n &= s_{n-2} e_{n-1} s_{n-2}.
\end{aligned} \tag{47.1}$$

These forms were motivated by the realization that all rank $n-1$ idempotents have the same path structure — each is conjugate to $(1](2)\cdots(n)$. For instance, $e_2 = s_1^{-1} e_1 s_1$ corresponds to the conjugate $(1)(2](3)\cdots(n)$ of $(1](2)\cdots(n)$, and is motivated by

$$(1)(2](3)\cdots(n) = (321) \circ (1](2)\cdots(n) \circ (123).$$

Similarly, $e_3 = s^{-1} e_2 s_1$ corresponds to the conjugate $(1)(2)(3](4)\cdots(n)$ of $(1)(2](3)\cdots(n)$, i.e.,

$$(1)(2)(3](4)\cdots(n) = (321) \circ (1)(2](3)\cdots(n) \circ (123).$$

Other σ-equivalent words that we shall need are itemized in Lemma 47.2 below. These equivalencies may be understood (on first pass) by placing them in the C_n context.

47.2 Lemma

Let $n \ge 3$, and let M_X be the free inverse monoid generated by $X = \{s_1, \dots, s_{n-2}, e\}$. Let σ be the congruence on M_X generated by the relations given in (46.2)*. Then when "equality" is understood to mean "equality modulo σ," we have the following:*

1. *Each $e_k\sigma$, $1 \le k \le n$, is an idempotent in M_X/σ.*
2. *In addition to the relations in* (46.2)*, we have relations $(es_1)^3 = (s_1e)^3 = (s_1e)^4$. Also, $(es_1)^3 = (s_1^{-1}e)^3 = (s_1^{-1}e)^4 = (es_1^{-1})^3 = (es_1^{-1})^4$.*

(3) *For relations among the* e_k *and the* s_i, *we have*

for $k = 1$, $\quad e_1 s_i = s_i e_2,\ i \geq 1,$ *and* $e_1 s_1^{-1} = s_1^{-1} e_3;$

for $k = 2$, $\quad e_2 s_i = s_i e_1,\ i \geq 2,\ e_2 s_1^{-1} = s_1^{-1} e_1,$ *and* $e_2 s_1 = s_1 e_3;$

for $k = 3$, $\quad e_3 s_i = s_i e_3,\ i \geq 3,\ e_3 s_1^{-1} = s_1^{-1} e_2,\ e_3 s_1 = s_1 e_1,$ *and* $e_3 s_2 = s_2 e_4;$

for $k \geq 4$, $\quad e_{k+1} s_{k-1} = s_{k-1} e_k$ *and* $e_k s_{k-1} = s_{k-1} e_{k+1};$ $e_k s_i = s_i e_k$ *where* $i \neq k-1, k-2;$ $e_k s_1^{-1} = s_1^{-1} e_k.$

(4) *For any word* w *in* M_X, *we have* $w = s_{j_1} s_{j_2} \cdots s_{j_q} e_{i_1} e_{i_2} \cdots e_{i_k}$.

(5) *For* s_1 *and* e_1, e_2, e_3, *we have the relations*

$$s_1 e_1 e_2 e_3 = e_1 e_2 e_3 s_1 = e_1 e_2 e_3 \quad \textit{and} \quad s_1^{-1} e_1 e_2 e_3 = e_1 e_2 e_3 s_1^{-1} = e_1 e_2 e_3.$$

Proof. (Of (1)): Use induction on k, $1 \leq k \leq n$: Clearly, $e_1\sigma = e\sigma$ is an idempotent since $(e^2, e) \in \sigma$. For $k > 1$, calculate $e_k e_k$ in terms of e_{k-1}, observing that $e_k\sigma$ is an idempotent if $e_{k-1}\sigma$ is. (Of (2)): To simplify notation, let $s = s_1$. Then, the first equality holds since

$$\begin{aligned}(es)^3 &= (ss^{-1})(es)^3 = s(s^{-1}es)eses = se_2e_1ses \\ &= se(s^{-1}es)ses = ses^{-1}e(s^{-1}es) && (\text{since } s^2 = s^{-1}) \\ &= ses^{-1}(s^{-1}es)e = sesese && (\text{since } s^{-2} = s).\end{aligned}$$

Using $(se)^3 = (es)^3$ and $(es)^3 = (es)^4$, the second equality will follow if

$$(es)^4 = (se)^4. \tag{47.3}$$

To see that (47.3) holds, use the fact that es commutes with $(es)^3 = (se)^3$:

$$(es)^4 = es(se)^3 = es(se)^3e = \big((se)^3es\big)e = (se)^4.$$

The equality $(es)^3 = (s^{-1}e)^3$ holds since $(es)^3\sigma$ is an idempotent:

$$\begin{aligned}(es)^3(es)^3 &= (es)^4(es)^2 = (es)^3(es)^2 \\ &= (es)^4(es) = (es)^3(es) = (es)^4 = (es)^3,\end{aligned}$$

and $(s^{-1}e)^3\sigma$ is the inverse of (and therefore equal to) $(es)^3\sigma$. The equality $(s^{-1}e)^3 = (s^{-1}e)^4$ follows since $(es)^3 = (es)^4$ and inverses are unique. Similarly, $(s^{-1}e)^3 = (es^{-1})^3$ and $(s^{-1}e)^4 = (es^{-1})^4$ since $(es)^3 = (se)^3$ and $(es)^4 = (se)^4$. (Of (3)): (Continue the notation $s = s_1$.) For $k = 1$, the first set of equalities follows from $es_i = s_is^{-1}es$ and $e_2 = s^{-1}es$; while the other equality follows from $se_1s^{-1} = s^{-1}s^{-1}ess = s^{-1}e_2s = e_3$ and multiplication (on the left) by s^{-1}. For $k = 2$, the first set (of equalities) follows by uniqueness of the inverse of $e_1s_i = s_ie_2$ (since $i \geq 2 \Rightarrow s_i^2 = 1$); while the other equalities, those involving either s^{-1} or s, follow from the definitions of e_2 and e_3, respectively. For $k = 3$, let $i \geq 3$, and consider

$$\begin{aligned}
e_3s_i &= s^{-1}s^{-1}esss_i = sesss_i && \text{(since } s^{-2} = s\text{)}\\
&= ses(s_is^{-1}) && \text{(since } (s_1s_i)^2 = 1 \text{ for } i \geq 3\text{)}\\
&= s(es_i)s^{-1}s^{-1} = ss_ie_2s^{-1}s^{-1} && \text{(from case } k = 1\text{)}\\
&= s_is^{-1}e_2s^{-1}s^{-1} = s_is^{-1}e_2s = s_ie_3.
\end{aligned}$$

And for equalities involving either s^{-1} or s, use the uniqueness of the inverse of $e_2s_1 = s_1e_3$ and the uniqueness of the inverse of $e_1s_1^{-1} = s_1^{-1}e_3$. The equality $e_3s_2 = s_2e_4$ follows from the definition of e_4 and $s_2^2 = 1$. Next, let $k \geq 4$. Then first, $e_{k+1} = s_{k-1}e_ks_{k-1}$ and $s_{k-1}^2 = 1$ yield

$$e_{k+1}s_{k-1} = s_{k-1}e_k \quad \text{and } e_ks_{k-1} = s_{k-1}e_{k+1}.$$

Second, continuing with $k \geq 4$, we show $e_ks_i = s_ie_k$, for $i \neq k-1, k-2$, by considering three subcases, $i \geq k$, and $1 < i < k-2$, and $i = 1$: Assuming $i \geq k \geq 4$, suppose $e_{k-1}s_i = s_ie_{k-1}$ for $i \geq k-1 \geq 3$. Then we have

$$\begin{aligned}
e_ks_i = (s_{k-2}e_{k-1}s_{k-2})s_i &= s_{k-2}e_{k-1}s_is_{k-2} && \text{(since } i \geq k \Rightarrow (s_{k-2}s_i)^2 = 1\text{)}\\
&= s_{k-2}s_ie_{k-1}s_{k-2} && \text{(since } i \geq k-1 \geq 3\text{)}\\
&= s_ie_k.
\end{aligned}$$

Thus, in particular, $e_ks_k = s_ke_k$ for all $k \geq 3$. Assuming $1 < i < k-2$, we may then conclude that

$$\begin{aligned}
e_ks_i &= (s_{k-2}\cdots s_{i+1}s_ie_{i+1}s_is_{i+1}\cdots s_{k-2})s_i\\
&= \cdots s_{i+1}s_ie_{i+1}(s_is_{i+1}s_i)\cdots\\
&= \cdots s_{i+1}s_ie_{i+1}(s_{i+1}s_is_{i+1})\cdots && \text{(since } (s_is_{i+1})^3 = 1\text{)}\\
&= \cdots s_{i+1}s_i(s_{i+1}e_{i+1})s_is_{i+1}\cdots && \text{(since } i+1 \geq 3\text{)}\\
&= \cdots (s_is_{i+1}s_ie_{i+1}s_is_{i+1})\cdots\\
&= s_ie_k.
\end{aligned}$$

Assuming $i = 1$ (and still using $s = s_1$), we have $e_ks = (s_{k-2}\cdots e_3\cdots s_{k-2})s$. And since $s_qs = s^{-1}s_q$ and $s_qs^{-1} = ss_q$ when $q > 2$, we also have, for $k \geq 4$,

$$e_ks = \begin{cases} \cdots s_2s_1^{-1}e_2s_1s_2s\cdots, & \text{when } k \text{ even}\\ \cdots s_3s_2s_1^{-1}e_2s_1s_2s^{-1}s_3\cdots, & \text{when } k \text{ odd.} \end{cases}$$

When $k \geq 4$ is even, we continue with

$$\begin{aligned}
e_k s_1 &= \cdots s_2 s^{-1} e_2 (s s_2 s) \cdots \\
&= \cdots s_2 s^{-1} e_2 s_2 s^{-1} s_2 \cdots && (\text{since } s s_2 s = s_2 s^{-1} s_2) \\
&= \cdots s_2 s^{-1} (s_2 e_1) s^{-1} s_2 \cdots && (\text{since } e_2 s_2 = s_2 e_1) \\
&= \cdots s s_2 s e_1 s^{-1} s_2 \cdots \\
&= s(s_{k-2} \cdots s_2 s e_1 s^{-1} s_2 \cdots s_{k-2}) && (\text{since } k \text{ even}) \\
&= s(s_{k-2} \cdots s_2 s^{-2} e_1 s^2 s_2 \cdots s_{k-2}) \\
&= s_1 e_k.
\end{aligned}$$

For the case where $k \geq 4$ is odd, we shall need $e_4 s^{-1} = s^{-1} e_4$, which follows, since inverses are unique, from the known equality $e_4 s = s e_4$. So continuing, where $k \geq 4$ is odd, we have

$$\begin{aligned}
e_k s_1 &= \cdots s_3 s_2 (e_3 s_2) s^{-1} s_3 \cdots \\
&= \cdots s_3 s_2 (s_2 e_4) s^{-1} s_3 \cdots \\
&= \cdots s_3 s_2 s_2 s^{-1} e_4 s_3 \cdots && (\text{since } e_4 s^{-1} = s^{-1} e_4) \\
&= s(s_{k-2} \cdots s_3 e_4 s_3 \cdots s_{k-2}) && (\text{since } s_2^2 = 1 \text{ and } k \text{ odd}) \\
&= s_1 e_k.
\end{aligned}$$

Finally, we consider $e_k s_1^{-1} = s_1^{-1} e_k$, which follows from the uniqueness of the inverse of $e_k s_1 = s_1 e_k$.
(Of (4)): This follows from part (3) since we can "move the e_k's to the right."
(Of (5)): From the definitions of e_1, e_2, e_3 and part (2) we have

$$\begin{aligned}
s_1 e_1 e_2 e_3 &= se(s^{-1} es)(s^{-1} s^{-1} ess) = s(es^{-1} es^{-1} es^{-1}) \\
&= s(eseses) = (sesese)s = (sesesese)s \\
&= (es^{-1} es^{-1} es^{-1} es^{-1})s = (es^{-1} es^{-1} es^{-1}) es^{-1} s \\
&= (es^{-1} es^{-1} es^{-1})e = (es^{-1} e(ss^{-1}) s^{-1} ess)e \\
&= (e(s^{-1} es)(s^{-1} s^{-1} ess))e = e_1 e_2 e_3 e_1 = e_1 e_2 e_3.
\end{aligned}$$

To see that s_1 commutes with $e_1 e_2 e_3$, use part (3), $k = 1, 2, 3$, and calculate $s_1 e_1 e_2 e_3 = e_3 s_1 e_2 e_3 = e_3 e_1 s_1 e_3 = e_3 e_1 e_2 s_1$. Thus, the first string of equalities in part (5) holds. The last string of equalities now follows by taking the inverse of each "word," i.e., σ-class, in the first string. □

§48 A_n^c Inductive Lemma

For $X = \{s_1, \ldots, s_{n-2}, e\}$ and for $\iota : X \to M_X/\sigma$ given by $x\iota = [x] = x\sigma \in M_X/\sigma$, we have $(M_X/\sigma, \iota)$ as the freest inverse monoid that is generated by

$X = \{s_1, \dots, s_{n-2}, e\}$ and that satisfies the relations (46.2). We also have $\phi : X \to A_n^c$ where $s_1\phi = (1,2,3)$, $s_i\phi = (12)(i+1, i+2)$ for $2 \leq i \leq n-2$, and $e\phi = (1](2)\cdots(n)$. The relative freeness of M_X/σ, coupled with the fact that $X\phi$ is a generating set of A_n^c, says that there is a unique epimorphism $\psi : M_X/\sigma \to A_n^c$ such that $(x\iota)\psi = (x\rho)\psi = x\phi$, i.e., the following diagram commutes:

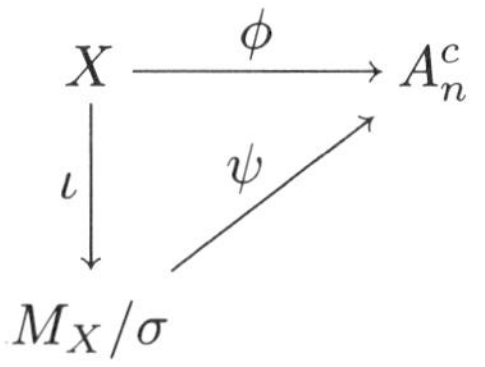

$$\begin{array}{ccc} X & \xrightarrow{\phi} & A_n^c \\ \iota \downarrow & \nearrow_{\psi} & \\ M_X/\sigma & & \end{array}$$

To simplify notation, we shall denote both ι (from X to M_X/σ) and the natural map $\sigma^\natural$ (from M_X to M_X/σ) as σ. Moreover, for any $B = \{i_1, i_2, \dots, i_m\} \subset \{1, 2, \dots, n\}$, let

$$e_B = e_{i_1} e_{i_2} \cdots e_{i_m}, \quad e_\emptyset = 1,$$

and then note that e_B is well-defined since idempotents in M_X/σ commute. Also define the composition $\sigma : M_X \to M_X/\sigma$ followed by ψ as the "hat" map $\widehat{\ } = \sigma \circ \psi$, i.e., $\widehat{w} = (w\sigma)\psi$. So when t is a product of elements of $\{s_1, \dots, s_{n-2}\}$, then $\widehat{te_B} = (te_B)\sigma\psi = (t)\sigma\psi\,(e_B)\sigma\psi = \widehat{t}\widehat{e}_B$.

48.1 Lemma

Let $t_{ijk} \in M_X$ be such that $\widehat{t}_{ijk} = (i,j,k) \in A_n$ is a 3-cycle. Then $t_{ijk}e_ie_je_k = e_ie_je_k$ (mod σ).

Proof. Certainly each of $\widehat{s}_1 = (1,2,3)$ and $\widehat{s_1^{-1}} = (1,3,2)$ is conjugate within the symmetric group S_n to $\widehat{t}_{ijk} = (i,j,k)$. If $\sigma^{-1}\widehat{s}_1\sigma = (i,j,k)$, where σ is odd, then

$$\big((2,3)\sigma\big)^{-1}\widehat{s_1^{-1}}\big((2,3)\sigma\big) = (i,j,k),$$

and $(2,3)\sigma$ is even. Thus, there exists $w \in M_X$ with $w\sigma \in \langle\{s_i\sigma\}\rangle$ and $\widehat{w} \in A_n$ such that either

$$\widehat{w}^{-1}\widehat{s}_1\widehat{w} = \widehat{t}_{ijk} \quad \text{or} \quad \widehat{w}^{-1}\widehat{s_1^{-1}}\widehat{w} = \widehat{t}_{ijk}.$$

Second, recall from the group theory case that ψ restricted to $\langle\{s_i\sigma\}\rangle$ is an isomorphism onto $A_n \subset A_n^c$. It then follows that either

$$w^{-1}s_1w = t_{ijk} \pmod{\sigma} \quad \text{or} \quad w^{-1}s_1^{-1}w = t_{ijk} \pmod{\sigma}.$$

(We shall assume the former σ-equivalence since the other case is similar). Third, we show that for each $m \in \{1,2,3\}$,

$$w^{-1}e_mw = e_{m\widehat{w}} \pmod{\sigma}. \tag{48.2}$$

To verify (48.2), apply (4) of 47.2 — "move e_m to the right," i.e.,

$$w^{-1}e_m w = ze_\ell \pmod{\sigma},$$

where $\widehat{z} \in A_n$. From this we deduce that $\widehat{w}^{-1}\widehat{e}_m\widehat{w} = \widehat{z}\widehat{e}_\ell$. On the other hand, it is clear that $\widehat{w}^{-1}\widehat{e}_m\widehat{w} = \widehat{e}_{m\widehat{w}}$, and so $\widehat{z}\widehat{e}_\ell = \widehat{e}_{m\widehat{w}}$. A comparison of the images of these two maps then gives $\ell = m\widehat{w}$, and it then follows that $i\widehat{z} = i$ except possibly for $i = m\widehat{w}$. Since $\widehat{z}$ is a permutation, we must have $\widehat{z} = 1$. Since the restriction of ψ to $\langle\{s_i\sigma\}\rangle$ is an isomorphism, $z = 1 \pmod{\sigma}$, and (48.2) now follows. From (48.2) and $w\sigma \in \langle\{s_i\sigma\}\rangle$, we have $w^{-1}e_1e_2e_3w = e_ie_je_k \pmod{\sigma}$. We can now calculate that $(w^{-1}s_1w)(w^{-1}e_1e_2e_3w) = t_{ijk}e_ie_je_k \pmod{\sigma}$. And from this and (5) of 47.2, we are finished. □

48.3 Lemma

Let t and t_0 be products of elements in $\{s_1, \ldots, s_{n-2}\}$. Then $\widehat{t}\widehat{e}_B = \widehat{t}_0\widehat{e}_B$ implies $(te_B, t_0e_B) \in \sigma$.

Proof. Let p_m be the statement:

$$p_m \quad \equiv \quad \widehat{t}\widehat{e}_B = \widehat{t}_0\widehat{e}_B \quad \text{when} \quad |B| \le m \quad \Rightarrow \quad (te_B, t_0e_B) \in \sigma.$$

We show p_0, p_1, p_2 and (the implication) "$p_{m-1} \Rightarrow p_m$" are true. If $m = 0$, then $B = \emptyset$ and $(t, t_0) \in \sigma$ since, from group theory [10, p. 298], the homomorphism ψ carries the subgroup $\langle\{s_i\sigma\}\rangle$ of M_X/σ isomorphically onto the (alternating) subgroup A_n of A_n^c. If $m = 1$, then both charts $\widehat{t}\widehat{e}_i$ and $\widehat{t}_0\widehat{e}_i$ have rank $n-1$ and are one-one. It follows that $\widehat{t} = \widehat{t}_0$, and again from the group theory, $(t, t_0) \in \sigma$. Hence, $(te_i, t_0e_i) \in \sigma$. If $m = 2$, then

$$\widehat{t}\widehat{e}_i\widehat{e}_j = \widehat{t}_0\widehat{e}_i\widehat{e}_j$$

where the indices $i, j \notin \mathbf{r}(\widehat{t}\widehat{e}_i\widehat{e}_j) = \mathbf{r}(\widehat{t}_0\widehat{e}_i\widehat{e}_j)$. In this case we have $B = \{i, j\}$, and we need to show $(te_B, t_0e_B) \in \sigma$. Again, it suffices to show that $\widehat{t} = \widehat{t}_0$. And to see that these *even* permutations are equal, we first pick an even permutation $\alpha \in A_n$ such that

$$\widehat{t}\alpha = \widehat{t}_0.$$

Second, we show that α is the identity permutation: For $k \in N = \{1, 2, \ldots, n\}$, if $k\widehat{t} \notin B$, then $k\widehat{t}_0 \notin B$ and $k\widehat{t}\alpha = k\widehat{t}_0 = k\widehat{t}$, showing that α fixes each member of $N - B$. So $\alpha = 1$ or $\alpha = (ij)$. Since α is even, it cannot be the transposition (ij). Thus, $\widehat{t} = \widehat{t}_0$.

Next suppose $m = |B| \ge 3$, that p_{m-1} is true, and that $\beta = \widehat{t}\widehat{e}_B = \widehat{t}_0\widehat{e}_B$. If, for every $l \in \{1, 2, \ldots, n\}$ $k\widehat{t} = x\widehat{t}_0$ then, again by the group theory case, we are finished. Otherwise, there is an $l \notin \mathbf{d}(\beta)$ such that

$$(48.4) \qquad j = l\widehat{t} \ne l\widehat{t}_0 = i \quad \text{and} \quad i, j \in B.$$

Since $|B| \geq 3$, we have the representation $\beta = \widehat{t}\widehat{e}_B = \widehat{t}\widehat{e}_i\widehat{e}_j\widehat{e}_k\widehat{e}_L = \widehat{t}_0\widehat{e}_i\widehat{e}_j\widehat{e}_k\widehat{e}_L = \widehat{t}_0\widehat{e}_B$, where $B = L \cup \{i, j, k\}$, $L \cap \{i, j, k\} = \emptyset$, and (48.4) holds. Now let $\widehat{t}_{ijk} = (i, j, k)$, and apply 48.1 to obtain

$$t_0 e_B = t_0 e_i e_j e_k e_L = t_0 t_{ijk} e_i e_j e_k e_L \pmod{\sigma}.$$

It follows that

$$t_* = t_0 t_{ijk} \quad \Rightarrow \quad t_0 e_B = t_* e_B \pmod{\sigma}.$$

We now observe that $\widehat{t}$ agrees with $\widehat{t}_*$ on $\mathbf{d}(\widehat{t}\widehat{e}_B) \cup \{x\}$. Thus, if $C = B - \{j\}$, we then have

$$\widehat{t}\widehat{e}_C = \widehat{t}_*\widehat{e}_C$$

where $|C| = m - 1$. Hence, by the inductive hypothesis, $(te_C, t_*e_C) \in \sigma$, showing that

$$(te_C e_j, t_* e_C e_j) = (te_B, t_* e_B) \in \sigma.$$

Since $t_0 e_B = t_* e_B \pmod{\sigma}$ we are finished. □

48.5 Theorem

Let $n \geq 3$, and let M_X be the free inverse monoid generated by the set $X = \{s_1, \ldots, s_{n-2}, e\}$. Let σ be the congruence on M_X generated by the following relations

$$\begin{aligned} s_1^3 = (s_i)^2 = (s_{i-1}s_i)^3 = (s_j s_k)^2 &= 1, && \textit{for} \quad i > 1, \ \ |j - k| > 1, \\ e^2 = e, \ \ (es_1)^3 &= (es_1)^4, && \textit{and} \\ (es_j)^2 = (es_j)^4, \ \ es_i &= s_i s_1^{-1} e s_1, && \textit{for} \quad j > 1, i \geq 1. \end{aligned}$$

Then M_X/σ is isomorphic to A_n^c.

Proof. From 47.2, we may assume that every $w \in M_X$ is in a σ-class with a word of the form te_B. So, to assume that ψ is not one-one is to contradict 48.3. We conclude that the epimorphism ψ is also a monomorphism. □

§49 Comments

Our approach to constructing defining relations for A_n^c runs parallel to that of C_n (Chapter 9). The common elements indicate a general method for finding presentations of certain inverse semigroups. To be more precise, let $G \subset S_n$ be a permutation group with presentation $\langle X \mid \Delta \rangle$, and consider the inverse semigroup $S = \{\alpha \in C_n \,|\, \alpha \text{ is a restriction of some } \gamma \in G\}$. Then what can be said about adding generators ($e^2 = e$) to obtain a presentation of the semigroup S?

In both of the A_n^c and C_n cases, generators $x_1, \ldots, x_k$ in the group presentation were first identified with permutations in cycle notation. These generators were then augmented with an idempotent generator $e = e^2$, which would ultimately correspond to the chart $\varepsilon = (1](2)\cdots(n)$. Then, using the fact that all idempotents of rank $n-1$ are conjugate, we "mechanically generated" new relations (between e and the x_i) by inspecting the path structure resulting from various multiplications of the charts corresponding to $x_1, \ldots, x_k, e$.

How do we know which relations to incorporate? Examples in the A_n^c case illustrate one answer. Consider the relations $(es_1)^3 = (es_1)^4$ and $(es_j)^2 = (es_j)^4$, $j > 1$. We calculate in path notation that es_1 must correspond to the chart with path form

$$(1](2)\cdots(n) \circ (1,2,3) = (2,3,1](4)\cdots(n).$$

Such a chart yields a cyclic semigroup of index 3 and period 1 (6.2). That is, the period 1 yields the exponent 4 on $(es_1)^4$. In contrast, the chart that corresponds to es_j, where $j > 1$, has path structure

$$(1](2)\cdots(n) \circ (1,2)(j+1,j+2) = (2,1](j+1,j+2)(k_1)\cdots(k_{n-4}).$$

In this case, we have a cyclic semigroup of index 2 and period 2, showing why we have equality $(es_j)^2 = (es_j)^{2+2}$ when $j > 1$. The idea is to pick those relations that show how the idempotent generator e "interacts" with each generator $x \in X$.

Having expanded the generating set X to $X \cup \{e\}$, and the set of relations Δ to a set $\Delta \cup \Delta_1$, we then encountered the word problem — we needed to establish that the unique morphism ψ was one-one. To do this, a technical lemma was needed to obtain canonical forms for words, and an inductive lemma was also needed to make use of the group presentation as the initial case in an inductive argument. (Proposition 42.11 played a key role in allowing for the use of the group presentation.)

The presentation of A_n used in this chapter was first introduced by Moore [1] and later discussed in Burnside [1]. The presentation of A_n^c first appeared in a paper written by the author [6] and communicated by J. M. Howie. During the review of that paper, Howie made several comments that led to a considerably improved proof of Lemma 48.1.

CHAPTER 11

Decomposing Partial Transformations

Path notation for charts (partial one-one transformations) is an extension of cycle notation for permutations. By allowing for the use of a right bracket "]" in conjunction with the usual parentheses "(" ")", path notation provides to charts what cycle notation provides to permutations. Here we take the next step, i.e., by allowing for the use of yet another symbol "$\rangle$", path notation is extended to partial transformations. In particular, partial transformations are decomposed into circuits and proper paths. The representation is then used to study idempotents, nilpotents, and cyclic semigroups.

§50 Semigroup Hierarchy

The *semigroup PT_n of partial transformations on* $N = \{1, 2, \ldots, n\}$ is the set of all functions $\alpha : \mathbf{d}\alpha \to \mathbf{r}\alpha$ that have domain $\mathbf{d}\alpha \subset N$ and range $\mathbf{r}\alpha \subset N$ under composition. Relative to S_n and C_n, the useful semigroup hierarchy is

$$S_n \subset C_n \subset PT_n \subset B_n,$$

where B_n is the semigroup of all binary relations $\alpha \subset N \times N$ under composition. In extending path notation from C_n to PT_n, we shall introduce the right angle "$\rangle$" notation, a notation that identifies those points where certain proper paths meet a circuit. Examples are provided in Figure 50.1, where members of PT_n are pictured geometrically.

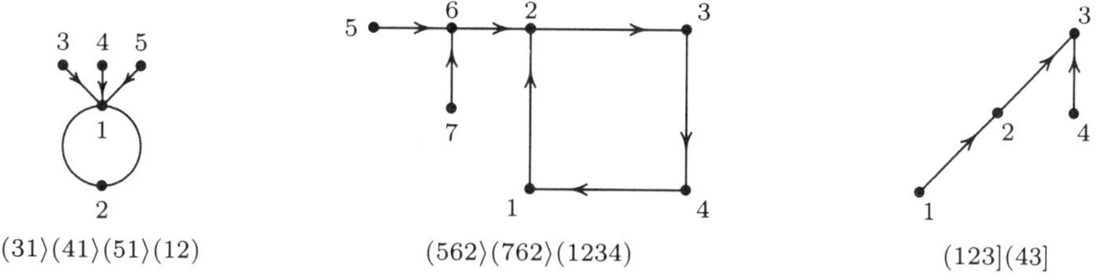

Figure 50.1. Partial transformations and path notation.

§51 Path Notation for Partial Transformations

Since $S_n \subset C_n \subset PT_n \subset B_n$, it is natural to extend the idea of "join in C_n" to *join* in B_n: For $\alpha, \beta \in B_n$, define the *join* $\alpha\beta$ as the union $\alpha \cup \beta$. In particular,

$$\alpha\beta \in \begin{cases} C_n & \text{if } \alpha, \beta \in C_n \text{ and } (\mathbf{d}\alpha \cup \mathbf{r}\alpha) \cap (\mathbf{d}\beta \cup \mathbf{r}\beta) = \emptyset \\ PT_n & \text{if } \alpha, \beta \in PT_n \text{ and } x \in \mathbf{d}\alpha \cap \mathbf{d}\beta \Rightarrow x\alpha = x\beta. \end{cases}$$

With this join operation, we could start with proper paths and circuits in C_n and then build partial transformations. We begin in reverse, however, starting with $\alpha \in PT_n$ and then defining certain paths induced by α. First, for each $x \notin \mathbf{d}\alpha \cup \mathbf{r}\alpha$, we shall call the expression "$(x]$" a *maximal proper path in* α. And then for $x \in \mathbf{d}\alpha$ and $k \geq 1$, we let $\eta_x = (x, x\alpha, \ldots, x\alpha^k]$ when the set $\{x, x\alpha, x\alpha^2, \ldots, x\alpha^k\}$ has size $k+1$; and $\gamma_x = (x, x\alpha, \ldots, x\alpha^{k-1})$ when $\{x, x\alpha, \ldots, x\alpha^{k-1}, x\alpha^k\}$ has size k and $x\alpha^k = x$. We call η_x a *proper path in* α, and γ_x (whenever it exists) a *circuit in* α. Such a proper path η_x is also *maximal* if its left endpoint $x \in \mathbf{d}\alpha - \mathbf{r}\alpha$ and its right endpoint $x\alpha^k \in \mathbf{r}\alpha - \mathbf{d}\alpha$. So maximal proper paths in α come in two varieties — those of the η_x kind and those expressions "$(x]$" where we have $x \notin \mathbf{d}\alpha \cup \mathbf{r}\alpha$.

For paths η and γ in α, we say that η *meets* γ whenever they are not disjoint (as charts). In particular, if $(\mathbf{d}\eta \cup \mathbf{r}\eta) \cap (\mathbf{d}\gamma \cup \mathbf{r}\gamma) = \{y\}$, then η *meets* γ at y; and if both η and γ are proper paths with a common proper terminal segment σ, we say that η *meets* γ *in* σ. To illustrate these concepts, consider the partial transformation

$$\alpha = \begin{pmatrix} 1 & 2 & 3 & 4 & 5 & 6 & 7 \\ 2 & 1 & 1 & 3 & 3 & 7 & - \end{pmatrix} \in PT_7$$

as pictured in Figure 51.1. Note that each of $\eta = (431]$, $\eta' = (531]$, and $\eta'' = (67]$ is a proper path in α, but that only η'' is maximal. Moreover, observe that $\gamma = (12)$ is a circuit in α, that both η and η' meet γ at 1, and that η meets η' in the common terminal segment $\sigma = (31]$.

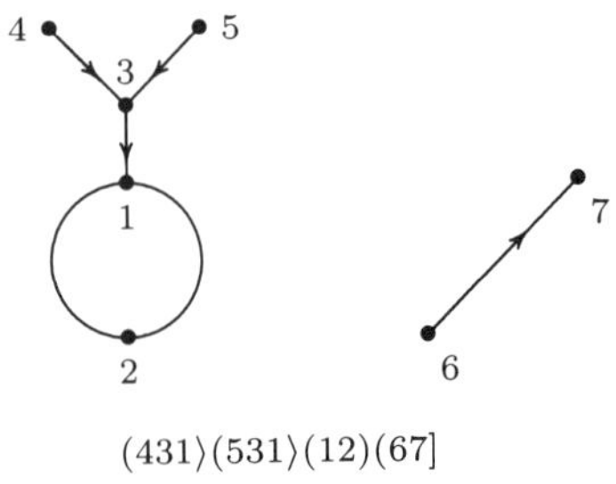

Figure 51.1. Maximal proper paths, circuits, and terminal segments.

51.2 Lemma

Let $\alpha \in PT_n$. Let η and η' be maximal proper paths in α, and let γ and γ' be circuits in α. Then the following statements are true:

(1) *If η meets η', then either $\eta = \eta'$ or η meets η' in a common proper terminal segment.*
(2) *Either $\gamma = \gamma'$ or γ does not meet γ'.*
(3) *For each $y \in \mathbf{r}\alpha - \mathbf{d}\alpha$ there exist $x \in \mathbf{d}\alpha - \mathbf{r}\alpha$ and $k \geq 1$ such that $x\alpha^k = y$, i.e., a maximal $\eta_x = (x, x\alpha, \dots, x\alpha^k = y]$ exists whenever $y \in \mathbf{r}\alpha - \mathbf{d}\alpha$.*

Proof. Statements (1) and (2) follow since $\alpha : \mathbf{d}\alpha \to \mathbf{r}\alpha$ is a function. For (3), we use induction: Since $y \in \mathbf{r}\alpha - \mathbf{d}\alpha$, there exists $x_1 \in \mathbf{d}\alpha$ such that $x_1\alpha = y$ and $\{x_1, y\}$ has size two. If $x_1 \notin \mathbf{r}\alpha$, then, letting $x = x_1$ and $k = 1$, we see that $\eta_x = (x_1, y] = (x, x\alpha^1 = y]$ is maximal. Otherwise, there exists $x_2 \in \mathbf{d}\alpha$ such that $x_2\alpha = x_1$ and $\{x_2, x_1, y\}$ has size three. If $x_2 \notin \mathbf{r}\alpha$, then, letting $x = x_2$ and $k = 2$, we see that $\eta_x = (x_2, x_1, y] = (x, x\alpha^1, x\alpha^2 = y]$ is maximal. But construction (by induction) of such sets $\{x_k, x_{k-1}, \dots, x_1, y\}$ must terminate since N is finite. It follows that the desired x and k exist. □

51.3 Theorem (Unique Representation of Partial Transformations)

Every transformation $\alpha \in PT_n - \{0\}$ is a join $\eta_1 \cdots \eta_u \gamma_1 \cdots \gamma_v$ of some (possibly none) length ≥ 2 proper paths $\eta_1, \dots, \eta_u$ and some (possibly none) circuits $\gamma_1, \dots, \gamma_v$ such that for indices i, j (distinct in (1) and (2)):

(1) *η_i meets η_j, if at all, in a common proper terminal segment;*
(2) *γ_i does not meet γ_j; and*
(3) *η_i meets γ_j, if at all, at the right endpoint of η_i.*

Moreover, this factorization is unique except for the order in which the paths are written.

Proof. Suppose $\alpha \in PT_n - \{0\}$. First, if $|\mathbf{d}\alpha - \mathbf{r}\alpha| = 0$, then α is a permutation of $\mathbf{d}\alpha$ and, consequently, a join of disjoint circuits $\gamma_1 \cdots \gamma_v$. Next, without loss of generality, suppose $\mathbf{d}\alpha - \mathbf{r}\alpha = \{1, 2, \dots, u\}$. Pick $x \in \mathbf{d}\alpha - \mathbf{r}\alpha$ and iterate, as far as possible, to obtain the set $X = \{x, x\alpha, x\alpha^2, \dots\}$. Since $X \subset N$ is finite, there exists a minimum $k \geq 1$ such that either $x\alpha^k \notin \mathbf{d}\alpha$ or $x\alpha^k = x\alpha^m$ for minimum $m > k$. In either case, define the proper path $\eta_x = (x, x\alpha, \dots, x\alpha^k]$. And in the case where the minimum $m > k$ exists, also define the circuit $\gamma_x = (x\alpha^k, \dots, x\alpha^{m-1})$. Supposing, again without loss of generality, that it is precisely when $x \in \{1, 2, \dots, w\} \subset \{1, 2, \dots, u\}$ that we obtain circuits γ_x, we define $\beta = \eta_1\eta_2 \cdots \eta_u\gamma_1 \cdots \gamma_w$. That these proper paths and circuits satisfy (1) – (3) follows from the definition of β and (1) – (2) of 51.2. Therefore, if $\alpha = \beta$ we are finished. Otherwise, we shall show that a join of circuits $\alpha' = \gamma_{w+1} \cdots \gamma_v$ exists such that $\alpha = \beta\alpha'$: First, define

$$\alpha' = \alpha|_{\mathbf{d}\alpha - \mathbf{d}\beta}$$

as the restriction of α to $\mathbf{d}\alpha - \mathbf{d}\beta$. To see that α' is a permutation, we shall need several properties:

(i) $\mathbf{r}\alpha - \mathbf{d}\alpha = \mathbf{r}\beta - \mathbf{d}\beta$ (Using (3) of 51.2, we have $y \in \mathbf{r}\alpha - \mathbf{d}\alpha \Leftrightarrow$ maximal $\eta_x = (x, \ldots, y]$ exists for some $x \in \mathbf{d}\alpha - \mathbf{r}\alpha \Leftrightarrow y \in \mathbf{r}\beta - \mathbf{d}\beta$.)
(ii) $\mathbf{d}\beta - \mathbf{r}\beta = \mathbf{d}\alpha - \mathbf{r}\alpha$ (Both equal $\{1, 2, \ldots u\}$.)
(iii) $\mathbf{r}\alpha' \cup \mathbf{r}\beta = \mathbf{r}\alpha$.

We can now show that $\mathbf{d}\alpha' \subset \mathbf{r}\alpha'$, implying that α' is a permutation of its domain $\mathbf{d}\alpha'$. Let $y \in \mathbf{d}\alpha'$. Then $y \in \mathbf{d}\alpha$ implies $y \notin \mathbf{r}\alpha - \mathbf{d}\alpha$. Coupling this with $y \notin \mathbf{d}\beta$ and (i), we deduce

(iv) $y \notin \mathbf{r}\beta$.

Similarly, from $y \notin \mathbf{d}\beta - \mathbf{r}\beta$ and $y \in \mathbf{d}\alpha$, (ii) shows

(v) $y \in \mathbf{r}\alpha$.

Then from (iv) and (v), property (iii) yields $y \in \mathbf{r}\alpha'$, which finishes the proof that α' is a permutation. Thus as promised, $\alpha' = \gamma_{w+1} \cdots \gamma_v$ is a join of disjoint circuits. Lastly, to see that the paths in the factorization $\alpha = \beta\alpha' = \eta_1 \cdots \eta_u \gamma_1 \cdots \gamma_v$ satisfy (1) – (3), it suffices to observe that $\mathbf{r}\beta \cap \mathbf{r}\alpha' = \emptyset$. (Otherwise, $z \in \mathbf{r}\beta \cap \mathbf{r}\alpha' = \mathbf{r}\beta \cap \mathbf{d}\alpha' \subset \mathbf{r}\beta - \mathbf{d}\beta$, showing z in the right side of (i) and contradicting the fact that z cannot be in the left side of (i) — $z \in \mathbf{d}\alpha'$ implies $z \in \mathbf{d}\alpha$.)

Turning to uniqueness, we suppose that $\alpha = \eta_1 \cdots \eta_u \gamma_1 \cdots \gamma_v = \eta_1' \cdots \eta_w' \gamma_1' \cdots \gamma_x'$ has two such factorizations. If $u = 0$, then $\alpha = \gamma_1 \cdots \gamma_v = \gamma_1' \cdots \gamma_x'$ is a join of circuits and the uniqueness follows from 51.2. So suppose $u \geq 1$. We claim that $u = w$: First, for any proper path η of length ≥ 2 we have $\eta = (1, 2, \ldots, k] \Rightarrow \{1\} = \mathbf{d}\eta - \mathbf{r}\eta$. Therefore, (1) and (3) (applied to $\eta_1 \cdots \eta_u$) imply $\mathbf{d}\alpha - \mathbf{r}\alpha$ has u elements and (to $\eta_1' \cdots \eta_w'$) w elements, i.e., $u = w$. So, by rearrangement if necessary, for each i, we may assume that $i =$ (left endpoint of η_i) $=$ (left endpoint of η_i'). Next, we make the observation that for any proper path η in any such decomposition of α, we have (η meets no circuit in α) $\Leftrightarrow$ (η is maximal in α). An application of this general fact and agreement at left endpoints (of η_i and η_i') show that we have only two cases:

(vi) The right endpoints of both η_i and η_i' are in $\mathbf{r}\alpha - \mathbf{d}\alpha$.
(vii) There exists a circuit γ in α such that both η_i and η_i' meet γ at their right endpoints.

In either case, η_i and η_i' must also have the same right endpoint, yielding $\eta_i = \eta_i'$. It therefore only remains to show that we may "cancel" the proper paths, i.e., show that $\gamma_1 \cdots \gamma_v = \gamma_1' \cdots \gamma_x'$, which is case $u = 0$. To this end, assume that for some index i, $y \in \mathbf{d}\gamma_i$. Then $y \notin \mathbf{d}(\eta_1 \cdots \eta_u)$ by (3). This implies $y \in \mathbf{d}\gamma'_j$ for some index j and, consequently, $y\gamma_i = y\gamma'_j$, showing that the function $\gamma'_1 \cdots \gamma'_x$ is an extension of $\gamma_1 \cdots \gamma_v$. Using a similar argument, we can also show the latter is an extension of the former, thereby finishing the proof of this theorem. □

§52 Cilia and Cells of Partial Transformations

From Theorem 51.3, each nonzero $\alpha \in PT_n$ is a join

$$\alpha = (a_{11} \cdots a_{1k_1}] \cdots (a_{u1} \cdots a_{uk_u}](b_{11} \cdots b_{1m_1}) \cdots (b_{v1} \cdots b_{vm_v})$$

of proper paths (of length ≥ 2) and circuits that satisfy (1)–(3) in 51.3. To this join, then, we may join the proper 1-paths $(j_i]$, $j_i \notin \mathbf{d}\alpha \cup \mathbf{r}\alpha$, yielding

$$\alpha = (j_1] \cdots (j_\ell](a_{11} \cdots a_{1k_1}] \cdots (a_{u1} \cdots a_{uk_u}](b_{11} \cdots b_{1m_1}) \cdots (b_{v1} \cdots b_{vm_v}).$$

We shall refer to this unique representation as either the *path decomposition* or *join representation* of α. In particular, $\alpha = 0 \in PT_n$ has join representation $(1] \cdots (n]$, even though the zero transformation 0 of PT_n is excluded from 51.3.

In the join representation of a partial transformation, proper paths are of two kinds, namely, those that meet circuits and those that do not meet circuits. We call each of the former kind a *cilium* (plural = *cilia*). For example,

$$\alpha = (1, 2, \ldots, i, x_0](x_0, x_1, \ldots, x_{m-1}) \in PT_n$$

is a join of a cilium $(1, 2, \cdots, i, x_0]$ and a circuit, which we clearly mark by replacing the right bracket "]" with the right angle "$\rangle$", yielding

$$\alpha = (1, 2, \ldots, i, x_0\rangle(x_0, x_1, \ldots, x_{m-1}).$$

We say that $(x_0, x_1, \ldots, x_{m-1})$ is *associated with* $(1, 2, \ldots, i, x_0\rangle$, and that $(1, 2, \ldots, i, x_0\rangle$ is *associated with* $(x_0, x_1, \ldots, x_{m-1})$. We may, in fact, have any finite number of cilia $\eta_1, \ldots, \eta_k$ associated with one circuit γ, and, in such a case, the join $\eta_1 \cdots \eta_k\gamma$ is called a *cell*. A typical cell is pictured in Figure 52.1, where we see the partial transformation

$$\begin{pmatrix} 1 & 2 & 3 & 4 & 5 & 6 & 7 & 8 & 9 & 10 & 11 & 12 \\ 2 & 3 & - & 3 & 6 & 7 & 8 & 10 & 7 & 8 & 11 & - \end{pmatrix} \in PT_{12},$$

whose path decomposition is $(1, 2, 3](4, 3](5, 6, 7, 8\rangle(9, 7, 8\rangle(8, 10)(11)(12]$. It is clear that this partial transformation has two cells, one with two cilia and the other with none.

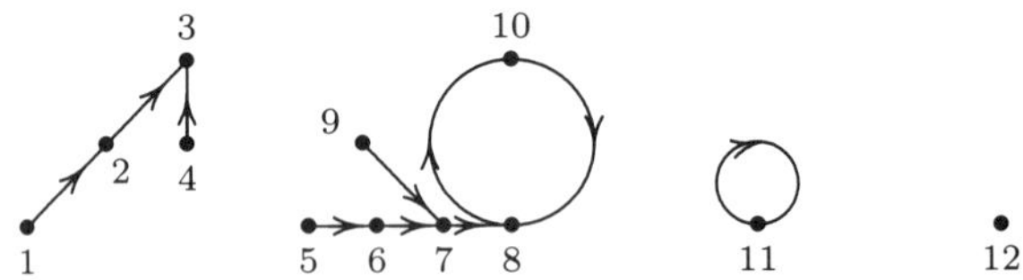

Figure 52.1. Decomposing a partial transformation.

§53 Idempotents and Nilpotents

In this section, we determine the path structure of both idempotent (53.1) and nilpotent (53.2) partial transformations. (The proof of the nilpotent case is obvious.)

53.1 Proposition

A partial transformation is an idempotent if and only if its path decomposition is a join of 1*-paths with some (possibly none) cilia of length* 2.

Proof. Suppose $\varepsilon^2 = \varepsilon$. If ε moves x to y, then ε must fix y. In other words, the path decomposition of ε must contain the cell $(xy\rangle(y)$. It follows that ε is a join of 1-paths with some (possibly none) cilia of length 2. The converse follows since any cell of the form $(x_1, y\rangle \cdots (x_k, y\rangle(y)$ is an idempotent. □

53.2 Proposition

A partial transformation is nilpotent if and only if its path decomposition contains no circuits.

To illustrate 53.1 and 53.2, consider $\varepsilon = (4](12\rangle(2)(35\rangle(5) \in PT_5$ and $\eta = (12](3](456](76] \in PT_7$.

§54 Multiplication of Partial Transformations

Learning to multiply partial transformations in path notation is like learning to multiply charts in path notation, it takes a little practice. When calculating powers α^k of a given $\alpha \in PT_n$, the next two lemmas are helpful. (Their proofs are straightforward.) In reading these lemmas, keep in mind that we are using "i" as a mnemonic for *index*.

54.1 Lemma

Let $\alpha = (12 \cdots i x_0\rangle(x_0 x_1 \cdots x_{m-1})$*, let* $k \geq i$*, and let* $\ell \in \{1, 2, \ldots, i\}$*. If we* *calculate subscripts modulo m, then*

$$\ell\alpha^k = \ell\alpha^{i-\ell+1}\alpha^{k-(i-\ell+1)} = x_0\alpha^{k-(i-\ell+1)} = x_{k-(i-\ell+1)}.$$

54.2 Lemma

Let the path decomposition of $\alpha \in PT_n$ be $\eta_1 \cdots \eta_u \gamma_1 \cdots \gamma_v (\eta_{11} \cdots \eta_{1m_1}\gamma'_1) \cdots (\eta_{w1} \cdots \eta_{wm_w}\gamma'_w)$ with each η_j a proper path, each γ_j a circuit, and each $(\eta_{j1} \cdots \eta_{jm_j}\gamma'_j)$ a cell. Then for each $k \geq 1$,

$$\alpha^k = \eta_1^k \cdots \eta_u^k \gamma_1^k \cdots \gamma_v^k (\eta_{11} \cdots \eta_{1m_1}\gamma'_1)^k \cdots (\eta_{w1} \cdots \eta_{wm_w}\gamma'_w)^k.$$

54.3 Theorem

If $\alpha \in PT_n$ has path decomposition $\alpha = (12 \cdots ix_0\rangle(x_0 \cdots x_{m-1})$, then the index of α is i.

Proof. From 54.1, for every $k \geq i$, we have $\alpha^k = (1, x_{k-i}\rangle(2, x_{k-(i-1)}\rangle \cdots (i, x_{k-1}\rangle(x_0 \cdots x_{m-1})^k$. On the other hand, if k is positive and less than i, then for $t = k - r \pmod m$ where $i = qk + r$ and $0 \leq r < k$, we have $1\alpha^k \in \{2, \ldots, i\}$, showing that $(1, 1\alpha^k, 1\alpha^{2k}, \ldots, x_t\rangle$ is a cilium of length at least 3 appearing in the decomposition of α^k. It then follows, since all cilia in the decompositions of powers of α greater than i have length 2, that the index of α cannot be less than i. To see that the index of α is i, note that $\alpha^i = (1, x_0\rangle(2, x_1\rangle \cdots (i, x_{i-1}\rangle(x_0 \cdots x_{m-1})^i$ and that

$$\alpha^i = \alpha^{i+m} = (1, x_{0+m}\rangle(2, x_{1+m}\rangle \cdots (i, x_{i-1+m}\rangle(x_0 \cdots x_{m-1})^{i+m},$$

which finishes the proof. □

54.5 Corollary

Let $\alpha = \eta_1 \cdots \eta_u \gamma_1 \cdots \gamma_v (\eta_{11} \cdots \eta_{1m_1}\gamma'_1) \cdots (\eta_{w1} \cdots \eta_{wm_w}\gamma'_w)$ be the decomposition of $\alpha \in PT_n$, with each η_j a proper path, each γ_j a circuit, and each $\eta_{j1} \cdots \eta_{jm_j}\gamma'_j$ a cell. Then for $\ell(\eta) =$ length of η,

$$\textit{index of } \alpha = \max\{\ell(\eta_1), \ldots, \ell(\eta_u), \ell(\eta_{11}) - 1, \ldots, \ell(\eta_{wm_w}) - 1\},$$

and the period of α is the least common multiple of the lengths of the circuits $\gamma_1, \ldots, \gamma_v, \gamma'_1, \ldots, \gamma'_w$. Moreover, if the join representation of α contains no proper paths (circuits), then the index (period) of α is 1.

Proof. Use 51.3, 54.2, 54.3, and the fact that the order of a permutation is the least common multiple of the lengths of its cycles. □

§55 Cyclic Semigroups of Partial Transformations

Expressing $\alpha \in PT_n$ in its join representation, we "know" (via 54.5) the index and period of $\langle\alpha\rangle$, i.e., we know $\langle\alpha\rangle$ up to isomorphism. As a consequence, the isomorphism classes of cyclic subsemigroups of PT_n may be determined by listing all possible forms of join representations.

Table 55.1. Index and Period for $\alpha \in PT_4 - C_4$.

number of $m \to k$ maps	**path structure**	**number**	**index**	**period**
$\binom{4}{2}\binom{4}{1}S(2,1)1! = 24$	$(21\rangle(1)(3](4]$	12	1	1
	$(13](23](4]$	12	2	1
$\binom{4}{3}\binom{4}{1}S(3,1)1! = 16$	$(21\rangle(31\rangle(1)(4]$	12	1	1
	$(14](24](34]$	4	2	1
$\binom{4}{4}\binom{4}{1}S(4,1)1! = 4$	$(21\rangle(31\rangle(41\rangle(1)$	4	1	1
$\binom{4}{3}\binom{4}{2}S(3,2)2! = 144$	$(1)(32\rangle(2)(4]$	24	1	1
	$(31\rangle(12)(4]$	24	1	2
	$(321\rangle(1)(4]$	24	2	1
	$(14](24](3)$	12	2	1
	$(14](234]$	24	2	1
	$(14](23\rangle(3)$	24	2	1
	$(124](324]$	12	3	1
$\binom{4}{4}\binom{4}{2}S(4,2)2! = 84$	$(2)(31\rangle(41\rangle(1)$	12	1	1
	$(21\rangle(1)(43\rangle(3)$	12	1	1
	$(41\rangle(23\rangle(31)$	12	1	2
	$(31\rangle(41\rangle(12)$	12	1	2
	$(321\rangle(41\rangle(1)$	24	2	1
	$(321\rangle(421\rangle(1)$	12	2	1
$\binom{4}{4}\binom{4}{3}S(4,3)3! = 144$	$(2)(3)(41\rangle(1)$	12	1	1
	$(2)(431\rangle(1)$	24	2	1
	$(23)(41\rangle(1)$	12	1	2
	$(1)(42\rangle(23)$	24	1	2
	$(431\rangle(12)$	24	2	2
	$(41\rangle(123)$	24	1	3
	$(4321\rangle(1)$	24	3	1
	Total =	**416**		

A listing of the possible forms of join representations in the $PT_4 - C_4$ case appears in Table 55.1. (Recall that the forms for charts in C_4 appear in Table 7.1.) The number of members of PT_4 is

$$1 + \sum_{1 \le k \le m \le 4} \binom{4}{m} \binom{4}{k} S(m,k)k! = 625.$$

In this formula, the Stirling number $S(m,k)$ of the second kind is the number of partitions of a set of m objects into k classes (see page 40 of Berge [1]). Among these 625 members of PT_4, we find

$$1 + \sum_{m=1}^{4} \binom{4}{m}^2 m! = 209$$

charts (partial one-one transformations), i.e., 209 of the 625 are members of $C_4 \subset PT_4$. Thus, our Table 55.1 displays the index and period of the $625 - 209 = 416$ members of $PT_4 - C_4$.

§56 Comments

According to Higgins [1, page 70], it was Suschkewitsch [3] who, in 1928, first depicted a full transformation $\alpha \in T_n \subset PT_n$ as a digraph on n vertices in which ij is a diedge whenever $i\alpha = j$. These digraphs are characterized by the condition that every vertex have out degree one, in which case they are called *functional digraphs*. The number of non-isomorphic functional digraphs having n vertices was determined in 1959 by Harary [1].

Since their introduction, many authors have used functional digraphs. For example, in 1968 Dénes [2] illustrated in his Figure 3 the concept of "main-permutation" of $\alpha \in T_n$, which, by considering $\alpha \in PT_n$, is the union of the circuits appearing in its path decomposition. And in 1988 these digraph representations (of full transformations) were employed by Higgins [2] in finding algorithms to solve equations such as

$$\alpha x^m \beta = \gamma \quad \text{and} \quad \alpha x = x\beta \quad (\alpha, \beta, \gamma \in T_n).$$

The approach taken in this chapter may be viewed as a variant of these digraph representations. Indeed, each $\alpha \in PT_n$ may also be pictured as a digraph on n vertices. In this case, however, the characterizing condition is that every vertex have out degree at most one. From this view, the path notation applied here to partial transformations captures the essential features of the digraph representation. The ideas and results of this chapter are due to Konieczny and Lipscomb [2].

CHAPTER 12

Commuting Partial Transformations

Like charts, commuting partial transformations $\alpha \circ \beta = \beta \circ \alpha$ also determine mappings of initial segments onto terminal segments and circuits onto circuits. In addition to circuits and segments, however, the PT_n case involves mappings of cilia. Exactly where a cilium in α may be mapped depends on whether or not its right-endpoint is in the domain of β.

We begin with examples of mappings of cilia, segments, and circuits (§57). Following the examples, we approach commutativity in PT_n by first representing $\alpha \in PT_n$ as the join $\alpha = \eta_1 \cdots \eta_u \gamma_1 \cdots \gamma_v$ of its maximal proper paths η_i and its cells γ_j (§58). Our goal (Theorem 58.8) is that of showing $\alpha \circ \beta = \beta \circ \alpha$ is equivalent to (*i*) β maps some (possibly none) initial segments of the η_i onto terminal segments of the η_i; (*ii*) β maps some (possibly none) of the cells γ_j onto subcells of the γ_j; and (*iii*) β maps some (possibly none) proper initial segments of the cilia onto terminal segments of the η_i.

An immediate application of this characterization yields the C_n case — since cells γ_j with no cilia are circuits, (*i*) and (*ii*) are the characterizing conditions for commuting charts. A more subtle application is that of computing the order $|C(\varepsilon)|$ of the centralizer $C(\varepsilon)$ of an idempotent $\varepsilon \in PT_n$ (§59).

§57 Examples

Let $\alpha = (123](45\rangle(567) = \eta\gamma$, where the maximal proper path $\eta = (123]$ and $(45\rangle$ is a cilium; and suppose that α and $\beta = (1)(2)(43\rangle(3)$ are partial transformations in PT_7 (Figure 57.1). Then $\alpha \circ \beta = (123] = \beta \circ \alpha$, and we see that β maps initial segments onto terminal segments, namely, $(123] \mapsto (123]$ and $(4] \mapsto (3]$. The former illustrates condition (*i*) above, while the latter illustrates (*iii*).

To illustrate condition (*ii*), consider $\alpha = (12](345\rangle(567) = \eta\gamma$, where the

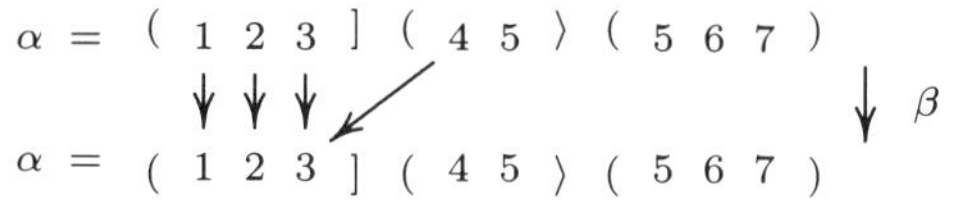

$$\begin{array}{l} \alpha = (\ 1\ 2\ 3\ ]\ (\ 4\ 5\ \rangle\ (\ 5\ 6\ 7\ ) \\ \qquad\ \downarrow\ \downarrow\ \downarrow\ \swarrow \qquad\qquad\qquad\qquad \downarrow \beta \\ \alpha = (\ 1\ 2\ 3\ ]\ (\ 4\ 5\ \rangle\ (\ 5\ 6\ 7\ ) \end{array}$$

Figure 57.1. Mapping initial segments onto terminal segments.

cell $\gamma = (345\rangle(567)$. Then $\sigma = (45\rangle(567)$ is a subcell of γ and, for the partial transformation $\beta = (345\rangle(567)$ illustrated in Figure 57.2, we see that β maps γ onto σ and that $\alpha \circ \beta = (35\rangle(576) = \beta \circ \alpha$.

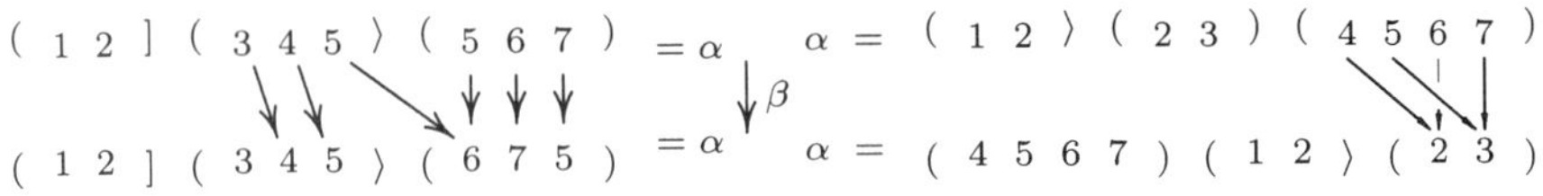

(1 2] (3 4 5 ⟩ (5 6 7) $= \alpha$ $\quad \downarrow \beta \quad$ $\alpha =$ (1 2 ⟩ (2 3) (4 5 6 7)

(1 2] (3 4 5 ⟩ (6 7 5) $= \alpha$ $\quad\quad\quad$ $\alpha =$ (4 5 6 7) (1 2 ⟩ (2 3)

Figure 57.2. Mapping cells onto subcells.

In the other picture in Figure 57.2, we have $\alpha = (12\rangle(23)(4567)$ and $\beta = (42](53](62](73]$. Again, β maps a cell (4567) onto a subcell (23) — (23) is a subcell of $(12\rangle(23)$ — and, as before, $\alpha \circ \beta = (43](52](63](72] = \beta \circ \alpha$. This example also illustrates the phenomenon of "folding" a k-circuit onto an ℓ-circuit $m = 2$ times, i.e., β folds the 4-circuit (4567) onto the 2-circuit (23) two times. It is obvious that such a folding can only occur when ℓ divides k (for $m > 2$, see Figure 58.4).

§58 Mapping Initial Segments and Cells

Recall that a partial transformation $\alpha \in PT_n$ may be expressed as a join $\alpha = \eta_1 \cdots \eta_u \gamma_1 \cdots \gamma_v$ of proper paths η_i (maximal in α) and cells $\gamma_j = \eta_{j1} \cdots \eta_{jm_j} \gamma_j'$, where the η_{jk} are the cilia associated with the circuit γ_j'. This representation allows us to describe those $\beta \in PT_n$ that commute with α — it is the action of β on the η_i, the η_{jk}, and the γ_j that determines whether β commutes with α. To deal with the η_i and the η_{jk}, we need to recall that (the representation) $(i_1 \cdots i_k]$ has

initial sets:	$\{i_1\}, \{i_1, i_2\}, \ldots, \{i_1, i_2, \ldots, i_k\}$;
terminal sets:	$\{i_1, i_2, \ldots, i_k\}, \ldots, \{i_{k-1}, i_k\}, \{i_k\}$;
initial segments:	$(i_1], (i_1 i_2], \ldots, (i_1 \cdots i_k]$; and
terminal segments:	$(i_1 \cdots i_k], \ldots, (i_{k-1} i_k], (i_k]$.

And for mapping segments onto segments, we extend the idea to PT_n. If β and the proper path $\eta = (i_1 \cdots i_k]$ are members of PT_n, and if there is an index w such that $\mathbf{d}\beta$ meets $\{i_1, \ldots, i_w, \ldots, i_k\}$ in the initial set $\{i_1, \ldots, i_w\}$, then for any proper path $\eta' = (j_1 \cdots j_v \ell_1 \cdots \ell_w]$ $(v \geq 0)$ such that $i_1\beta = \ell_1$, $\ldots$, $i_w\beta = \ell_w$, we shall say that β *maps an initial segment of* η *onto a terminal segment of* η'.

In the following lemma, we consider commuting partial transformations α and β. The focus is on how β maps proper paths in α.

58.1 Lemma (**Mapping initial segments onto terminal segments**)

Let $\alpha, \beta \in PT_n$ be such that $\alpha \circ \beta = \beta \circ \alpha$, and let η be a proper path $(i_1 \cdots i_k]$ in α such that some $i_u \in \mathbf{d}\beta$. If η is maximal in α, or if η is a cilium such that $i_k \notin \mathbf{d}\beta$, then β maps an initial segment of η onto a terminal segment of some maximal proper path in α.

Proof. Let $x, y \in N = \{1, \ldots, n\}$ and suppose $\delta \in PT_n$. We shall use a diagram $x \xrightarrow{\delta} y$ to mean that $x \in \mathbf{d}\delta$ and $x\delta = y$. The proof is based on the observation that $\alpha \circ \beta = \beta \circ \alpha$ is equivalent to the following two conditions:

(58.2) Any diagram
$$\begin{array}{ccc} x & \xrightarrow{\alpha} & y \\ & & \downarrow \beta \\ & & z \end{array} \quad \text{extends to} \quad \begin{array}{ccc} x & \xrightarrow{\alpha} & y \\ \beta \downarrow & & \downarrow \beta \\ w & \xrightarrow[\alpha]{} & z \end{array}$$

(58.3) Any diagram
$$\begin{array}{ccc} x & & \\ \beta \downarrow & & \\ w & \xrightarrow[\alpha]{} & z \end{array} \quad \text{extends to} \quad \begin{array}{ccc} x & \xrightarrow{\alpha} & y \\ \beta \downarrow & & \downarrow \beta \\ w & \xrightarrow[\alpha]{} & z \end{array}$$

So suppose $\alpha \circ \beta = \beta \circ \alpha$, i.e., that both (58.2) and (58.3) hold. Consider first the case where $\eta = (i_1 \cdots i_k]$ is maximal in α. If w is the largest index such that $i_w \in \mathbf{d}\beta$, then, by (58.2),

$$\begin{array}{ccc} i_{w-1} & \xrightarrow{\alpha} & i_w \\ & & \downarrow \beta \\ & & \ell_w \end{array} \quad \text{extends to} \quad \begin{array}{ccccccc} i_1 & \xrightarrow{\alpha} \cdots & \xrightarrow{\alpha} & i_{w-1} & \xrightarrow{\alpha} & i_w \\ \beta \downarrow & & & \beta \downarrow & & \downarrow \beta \\ \ell_1 & \xrightarrow[\alpha]{} \cdots & \xrightarrow[\alpha]{} & \ell_{w-1} & \xrightarrow[\alpha]{} & \ell_w \end{array}$$

By (58.3) and the maximality of w, we have $\ell_w \notin \mathbf{d}\alpha$, implying that ℓ_w is the right-endpoint of a maximal proper path $(j_1 \cdots j_v \ell_1 \cdots \ell_w](v \geq 0)$. It follows that we are finished with the case where η is maximal in α. So now suppose that $\eta = (i_1 \cdots i_k\rangle$ is a cilium in α with $i_k \notin \mathbf{d}\beta$. Again, letting w be the largest index such that $i_w \in \mathbf{d}\beta$, we obtain the diagram above. By (58.3), the maximality of w, and the fact that $i_k \notin \mathbf{d}\beta$, we have $\ell_w \notin \mathbf{d}\alpha$. The desired result follows. □

Turning to cells and subcells, we recall that when both α and β are charts, then each cell γ in α has no cilia, and that $\alpha \circ \beta = \beta \circ \alpha$ restricts β to mapping k-circuits onto k-circuits. When β is not a chart, however, β may "fold" a k-circuit onto an ℓ-circuit m times ($k = m\ell$). This folding is depicted in Figure 58.4, where the arrows are mappings of the unit circle S^1 (in the complex plane) onto itself, e.g., the $m = 2$ map $S^1 \to S^1$ given by $z \mapsto z^2$.

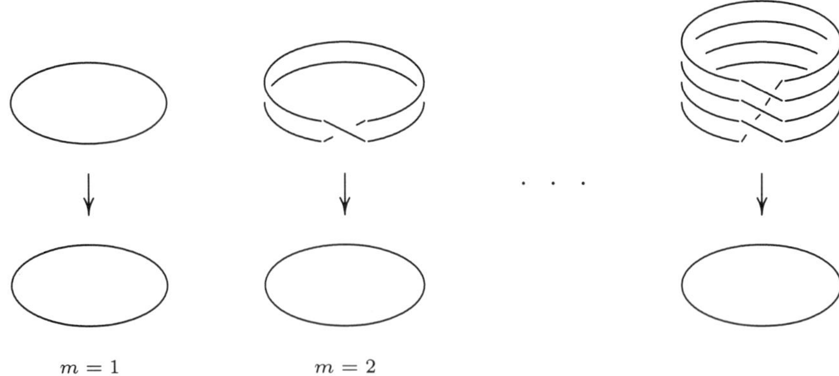

Figure 58.4. Folding a k-circuit onto an ℓ-circuit m times.

Even when the circuits have associated cilia, we may have foldings. For instance, by slightly modifying the pictures in Figure 58.4, we may fold both a cilium and its associated circuit onto a circuit (Figure 58.5).

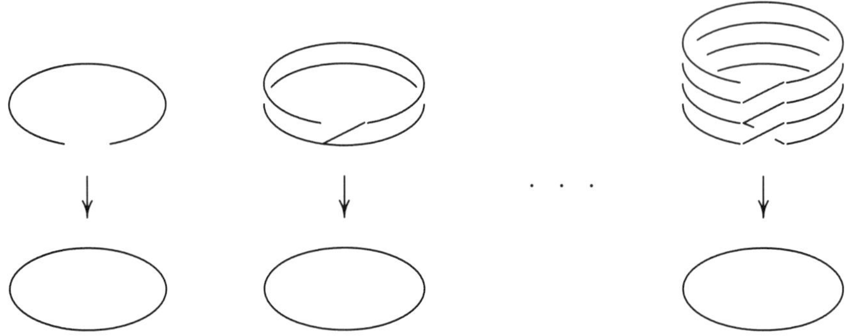

Figure 58.5. Folding both a cilium and its associated circuit onto a circuit.

To deal with these foldings, for any cell $\sigma = \eta_1 \cdots \eta_k \sigma'$ with cilia η_i and circuit σ', we define a *subcell* of σ to be any partial transformation $\eta_1' \cdots \eta_k' \sigma'$ where each η_i' is a terminal segment of the corresponding η_i. It follows that each subcell of σ, as a partial transformation, is also a cell; and under inclusion, σ and σ' are, respectively, the maximum and minimum subcells of σ. Subcells may be pictured as in Figure 58.7, where the circle representing σ' is oriented clockwise, and the edges that collectively represent the cilia are oriented "toward the circle."

In general, each cell corresponds to a digraph — one that has one circuit and all of its other diedges (defined by the cilia) directed toward the circuit. This digraph view of a cell motivates the following definition, where the subscript arithmetic is the appropriate modulo arithmetic.

58.6 Definition (**Morphisms of Cells**)

Let $\alpha, \beta \in PT_n$. Let α have a cell γ, with k-circuit $\gamma' = (i_0 \cdots i_{k-1})$, and a cell σ, with ℓ-circuit $\sigma' = (j_0 \cdots j_{\ell-1})$. We say that β maps γ onto a subcell of σ if ℓ divides k, if $\mathbf{d}\gamma \subset \mathbf{d}\beta$, if for some u, $i_0\beta = j_u$, $i_1\beta = j_{u+1}, \ldots$, and if for every cilium $\eta = (n_1 \cdots n_p\rangle$ in γ, exactly one of the following holds: There is a $j_w \in \mathbf{d}\sigma'$ such that $n_1\beta = j_w$, $n_2\beta = j_{w+1}$, $\ldots$; or there is a terminal segment $(m_1 \cdots m_t j_w\rangle$ of a cilium in σ such that $1 \leq t < p$ and

$$n_1\beta = m_1, \ \ldots\ , n_t\beta = m_t, \ n_{t+1}\beta = j_w, \ n_{t+2}\beta = j_{w+1}, \ \ldots \ .$$

If γ and σ are cells in α, and if β maps γ onto a subcell of σ, then it follows from 58.6 that α restricted to $(\mathbf{d}\gamma)\beta$ is a subcell of σ.

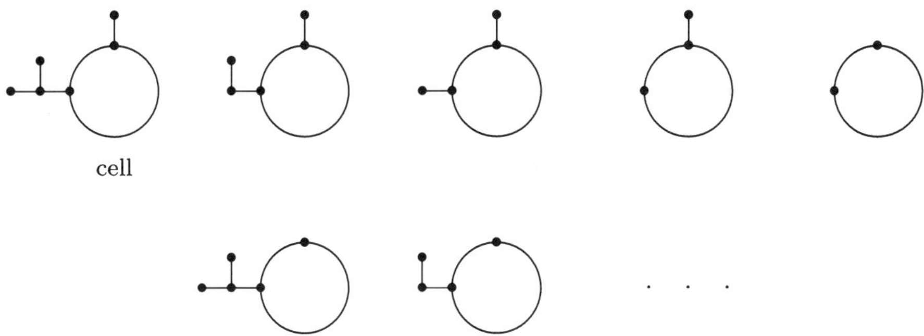

Figure 58.7. A cell and some of its subcells.

58.8 Theorem

Let $\alpha, \beta \in PT_n$. Then $\alpha \circ \beta = \beta \circ \alpha$ if and only if

(*i*) *when $\eta = (i_1 \cdots i_k]$ is maximal in α such that some $i_u \in \mathbf{d}\beta$, then β maps an initial segment of η onto a terminal segment of some maximal proper path in α;*

(*ii*) *when γ is a cell in α with circuit $\gamma' = (i_0 \cdots i_{k-1})$ such that some $i_u \in \mathbf{d}\beta$, then β maps γ onto a subcell of some cell in α; and*

(*iii*) *when $\gamma = \cdots \eta \cdots \gamma'$ is a cell in α with circuit $\gamma' = (i_0 \cdots i_{k-1})$ and a cilium $\eta = (\cdots w \cdots i_u\rangle$ such that $i_u \notin \mathbf{d}\beta$ but $w \in \mathbf{d}\beta$, then β maps an initial segment of η onto a terminal segment of some maximal proper path in α.*

Proof. Suppose $\alpha \circ \beta = \beta \circ \alpha$. Then (i) and (iii) follow from 58.1. To see that (ii) holds, we shall use 58.6 and the notation therein. So we already have $\gamma' = (i_0 \cdots i_{k-1})$ and some $i_u \in \mathbf{d}\beta$. Our first observation is that we may assume that $u = 0$. Second, we define $j_0 = i_0\beta$. Coupling this with the fact that γ' is a circuit in α, we have

$$\begin{array}{ccc} i_{k-1} & \xrightarrow{\alpha} & i_0 \\ & & \big\downarrow \beta \\ & & j_0, \end{array}$$

which, by (58.2), extends to the commutative diagram

$$\begin{array}{ccccccccc} i_0 & \xrightarrow{\alpha} & i_1 & \xrightarrow{\alpha} & \cdots & \xrightarrow{\alpha} & i_{k-1} & \xrightarrow{\alpha} & i_0 \\ \beta\big\downarrow & & \beta\big\downarrow & & & & \beta\big\downarrow & & \big\downarrow\beta \\ j_0 & \xrightarrow[\alpha]{} & j_1 & \xrightarrow[\alpha]{} & \cdots & \xrightarrow[\alpha]{} & j_{k-1} & \xrightarrow[\alpha]{} & j_0. \end{array}$$

It follows that β folds the k-circuit γ' onto an ℓ-circuit $\sigma' = (j_0 \cdots j_{\ell-1})$, where ℓ is either the smallest index in $\{1, \ldots, k-1\}$ such that $j_\ell = j_0$, or (if no such index exists) $\ell = k$. Moreover, we see that ℓ must divide k; and we define σ to be the cell in α that contains σ'. To see that $\mathbf{d}\gamma \subset \mathbf{d}\beta$, suppose $\eta = (n_1 \cdots n_p\rangle$ is associated with γ', then, from the diagram above, there is an index u such that $n_p\beta = j_u$. So again, we have

$$\begin{array}{ccc} n_{p-1} & \xrightarrow{\alpha} & n_p \\ & & \big\downarrow \beta \\ & & j_u, \end{array}$$

which, by (58.2), extends to the commutative diagram

$$\begin{array}{ccccccccc} n_1 & \xrightarrow{\alpha} & n_2 & \xrightarrow{\alpha} & \cdots & \xrightarrow{\alpha} & n_{p-1} & \xrightarrow{\alpha} & n_p \\ \beta\big\downarrow & & \beta\big\downarrow & & & & \beta\big\downarrow & & \big\downarrow\beta \\ k_1 & \xrightarrow[\alpha]{} & k_2 & \xrightarrow[\alpha]{} & \cdots & \xrightarrow[\alpha]{} & k_{p-1} & \xrightarrow[\alpha]{} & j_u. \end{array} \tag{58.9}$$

Thus, $\mathbf{d}\eta \subset \mathbf{d}\beta$, and it follows that $\mathbf{d}\gamma \subset \mathbf{d}\beta$. So at this point in the proof, we have $(n_1 \cdots n_p\rangle$ $(p > 1)$ in γ and need to address the cases $n_1\beta \in \mathbf{d}\sigma'$ and $n_1\beta \notin \mathbf{d}\sigma'$: If $n_1\beta \in \mathbf{d}\sigma'$, then for some index w, we have $n_1\beta = j_w$, and the diagram (58.9) becomes

$$\begin{array}{ccccccc} n_1 & \xrightarrow{\alpha} & n_2 & \xrightarrow{\alpha} & \cdots & \xrightarrow{\alpha} & n_p \\ \beta\big\downarrow & & \beta\big\downarrow & & & & \big\downarrow\beta \\ j_w & \xrightarrow[\alpha]{} & j_{w+1} & \xrightarrow[\alpha]{} & \cdots & \xrightarrow[\alpha]{} & j_u \end{array}$$

as specified in 58.6. In the other case, $n_1\beta \notin \mathbf{d}\sigma'$ and we first choose t as the largest index such that $n_t\beta \notin \mathbf{d}\sigma'$, which implies that $n_{t+1}\beta = j_w$ for some index w. Then the diagram (58.9) becomes

$$\begin{array}{ccccccccccc}
n_1 & \xrightarrow{\alpha} \cdots & \xrightarrow{\alpha} & n_t & \xrightarrow{\alpha} & n_{t+1} & \xrightarrow{\alpha} & n_{t+2} \cdots & \xrightarrow{\alpha} & n_p \\
\beta\downarrow & & & \beta\downarrow & & \beta\downarrow & & \downarrow\beta & & \downarrow\beta \\
m_1 & \xrightarrow[\alpha]{} \cdots & \xrightarrow[\alpha]{} & m_t & \xrightarrow[\alpha]{} & j_w & \xrightarrow[\alpha]{} & j_{w+1} \cdots & \xrightarrow[\alpha]{} & j_u
\end{array}$$

as specified in 58.6. It follows, since $m_t \notin \mathbf{d}\sigma'$, that the desired terminal segment $(m_1 \cdots m_t j_w\rangle$ of a cilium in σ exists. This finishes the proof of (ii). To see the converse, suppose that $\alpha, \beta \in PT_n$ satisfy (i), (ii), and (iii). We shall show that $\alpha \circ \beta = \beta \circ \alpha$ by showing α and β satisfy both (58.2) and (58.3). For (58.2), we begin with the diagram

$$\begin{array}{ccc}
x & \xrightarrow{\alpha} & y \\
 & & \beta\downarrow \\
 & & z.
\end{array}$$

Then $x \in \mathbf{d}\alpha$, which implies that x is either in a proper path $(\cdots xy \cdots]$ (maximal in α) or a cell in α. If $(\cdots xy \cdots]$ is maximal in α, then (i) and $y \in \mathbf{d}\beta$ imply that there is a maximal proper path $(\cdots x\beta\, y\beta \cdots]$ in α, in which case we are finished. On the other hand, if $(\cdots xy \cdots)$ is a circuit in α, then apply (ii). Finally, if $(\cdots xy \cdots t\rangle$ is a cilium in α, then apply (ii) when $t \in \mathbf{d}\beta$, and (iii) when $t \notin \mathbf{d}\beta$. Thus, (58.2) is true. The proof of (58.3) is similar. □

§59 Orders of Centralizers of Idempotents

From 53.1, an idempotent $\varepsilon \in PT_n$ is a join $\varepsilon = \eta_1 \cdots \eta_{m_0}\varepsilon_1 \cdots \varepsilon_p$ of proper 1-paths η_k and cells $\varepsilon_j = (l_{j1}, l_j\rangle(l_{j2}, l_j\rangle \cdots (l_{jm_j}, l_j\rangle(l_j)$ $(m_j \geq 0)$, where the cilia $(l_{jq}, l_j\rangle$ (if they exist) are 2-paths. We note that $m_j \geq 0$ denotes the number of cilia in ε_j, and, in particular, if $m_j = 0$, then $\varepsilon_j = (l_j)$ is a 1-circuit. We also observe that $m_0 + m_1 + \cdots + m_p$ counts the number of proper paths in the path decomposition of ε.

59.1 Proposition

Let $\varepsilon = \eta_1 \cdots \eta_{m_0}\varepsilon_1 \cdots \varepsilon_p$ be the join representation of an idempotent $\varepsilon \in PT_n$, where the η_k are maximal proper 1-paths and the ε_j are cells with $m_j \geq 0$ cilia. For $0 \leq i \leq p$ and $1 \leq j \leq p$, let

$$s_i = \sum_{k=0}^{m_i} \binom{m_i}{k} m_0^k \quad \text{and} \quad t_j = \sum_{i=1}^{p} (m_i + 1)^{m_j}.$$

Then

$$|C(\varepsilon)| = s_0(s_1+t_1)\cdots(s_p+t_p)$$

is the formula for the order $|C(\varepsilon)|$ *of the centralizer* $C(\varepsilon)$ *of* $\varepsilon \in PT_n$.

Proof. We first observe that s_0 is the number of ways to map (some) initial segments of the η_u onto terminal segments of the η_u; and, for $1 \le i \le p$, that the number s_i counts the ways to map (some) proper initial segments of cilia in ε_i onto terminal segments of the η_u. Moreover, for $1 \le j \le p$, we may also observe that the number t_j counts the ways to map the cell ε_j onto a subcell of the cell ε_i. These conclusions concerning the s_i and the t_j follow from the simple path structure of the η_u and the ε_v. As for the formula for $|C(\varepsilon)|$, note that if β commutes with ε, then β may be written as a join $\beta_0\cdots\beta_p$ where each $\beta_i \in PT_n$ with $\mathbf{d}\beta_0 \subset \{1,\dots,n\} - \mathbf{d}\varepsilon$ and $\mathbf{d}\beta_i \subset \mathbf{d}\varepsilon_i$ for $i \ge 1$. Furthermore, each of these induced β_i must also commute with ε. Conversely, the join β of any such list $\beta_0,\dots,\beta_p \in PT_n$ commutes with ε. Since the β_i may be counted independently, and since 58.8 shows that the count for β_0 is s_0, that the count for β_1 is (s_1+t_1), ..., and that the count for β_p is (s_p+t_p), the formula follows. □

To demonstrate 59.1, we shall consider an example in PT_7, which has $(7+1)^7 = 2,097,152$ members. Consider $\varepsilon = (1](2](3](45\rangle(65\rangle(5)(7) \in PT_7$. Then $p = 2$ and we have $m_0 = 3$, $m_1 = 2$, and $m_2 = 0$. Using the formula for the s_i and the t_j, we calculate that $s_0 = 64$, $s_1 = 16$, $s_2 = 1$, $t_1 = 10$, and $t_2 = 2$. It follows that

$$|C(\varepsilon)| = s_0\cdot(s_1+t_1)(s_2+t_2) = 64(16+10)(1+2) = 4992.$$

§60 Howie's Theorem

Recall that a mapping or function from N into N is called a *full transformation* of N; and that the subsemigroup of PT_n consisting of all full transformations of $N = \{1,\dots,n\}$ is the *full transformation semigroup* T_n. For $n \ge 2$, the semigroup T_n is not an inverse semigroup because its idempotents do not necessarily commute, e.g., $(12\rangle(2) \circ (21\rangle(1) = (21\rangle(1) \ne (12\rangle(2) = (21\rangle(1) \circ (12\rangle(2)$. But T_n is a regular semigroup: For $\alpha \in T_n$, let $\beta_1 : \mathbf{r}\alpha \to N$ be a choice function $y\beta_1 = x_y \in y\alpha^{-1}$. It easily follows that $y(\beta_1 \circ \alpha) = y$ for every $y \in \mathbf{r}\alpha$. Then, letting

$$\beta_2 = (\alpha \circ \beta_1)|_{N-\mathbf{r}\alpha},$$

we define β as the join of β_1 and β_2, i.e., $\beta = \beta_1\beta_2$. And since this β satisfies $\alpha \circ \beta \circ \alpha = \alpha$, it follows that T_n is regular. Moreover, recall that the *rank of* $\alpha \in T_n$ is the size $|\mathbf{r}\alpha|$ of its range $\mathbf{r}\alpha$. One important aspect of rank is that it determines $\mathcal{D}$ classes. In fact, recall that $\ker \alpha = \alpha \circ \alpha^{-1}$ where $\alpha^{-1} = \{ (x,y) \mid (y,x) \in \alpha \}$, and that Green's $\mathcal{R}$, $\mathcal{L}$, and $\mathcal{D}$ classes (in T_n) are determined by

$$\begin{aligned} \alpha \mathcal{R} \beta &\Leftrightarrow \alpha \circ \alpha^{-1} = \beta \circ \beta^{-1} \ (\ker \alpha = \ker \beta), \\ \alpha \mathcal{L} \beta &\Leftrightarrow \mathbf{r}\alpha = \mathbf{r}\beta, \quad \text{and} \\ \alpha \mathcal{D} \beta &\Leftrightarrow |\mathbf{r}\alpha| = |\mathbf{r}\beta|. \end{aligned}$$

Using these equivalences, we see the egg-box structure of T_1, T_2, and T_3 in Figure 60.1, where the elements in bold-face font are idempotents.

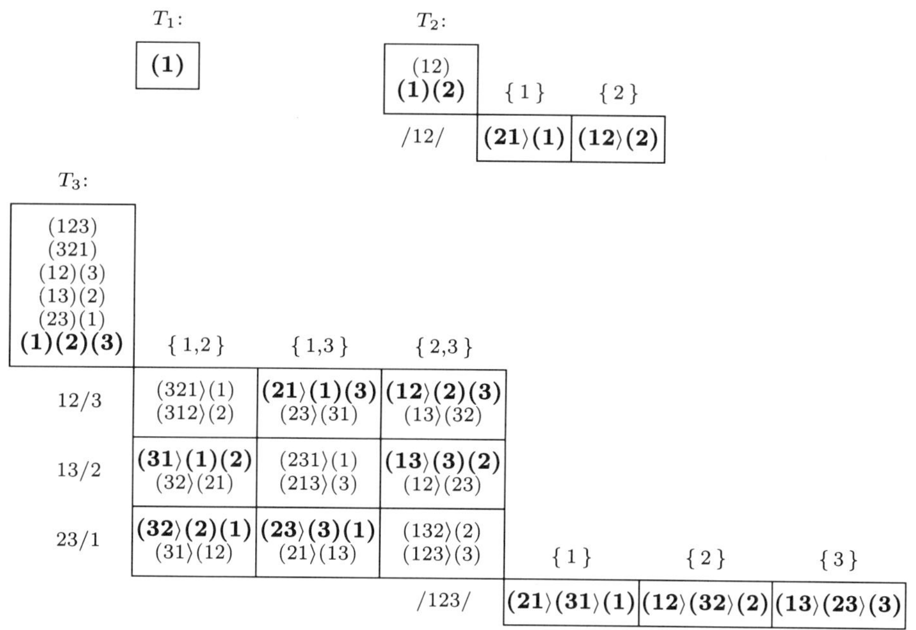

Figure 60.1. Egg-box structures of full transformation semigroups.

It is a trivial observation (Figure 60.1) that the idempotents of rank one generate $T_2 - S_2$, which brings us to Howie's theorem: Every element in $T_n - S_n$ is a product of rank $n-1$ idempotents in T_n. In other words, the semigroup S generated by the rank $n-1$ idempotents of T_n is $S = T_n - S_n$.

Note that each rank $n-1$ idempotent $\varepsilon \in T_n$ has path decomposition $\varepsilon = (xi_1\rangle(i_1)(i_2)\dots(i_{n-1})$, which we shall abbreviate as $\varepsilon = (xi_1\rangle(i_1)$. (In general, 1-circuits may be omitted.)

The proof of the following lemma is an exercise in path multiplication.

60.2 Lemma

For any $p \geq 2$ and $k \geq 2$:

(1) $(1 \cdots pi\rangle(i) = (pi\rangle(i) \circ (p-1, p\rangle(p) \circ \cdots \circ (12\rangle(2)$;
(2) $(1i_1\rangle(i_1 \cdots i_k) = (i_k 1\rangle(1) \circ (i_{k-1} i_k\rangle(i_k) \circ \cdots \circ (i_1 i_2\rangle(i_2) \circ (1i_1\rangle(i_1)$;
(3) $(1 \cdots pi_1\rangle(i_1 \ldots i_k) = (i_k p\rangle(p) \circ (i_{k-1} i_k\rangle(i_k) \circ \cdots \circ (i_1 i_2\rangle(i_2) \circ (pi_1\rangle(i_1)$
$\circ (p-1, p\rangle(p) \circ \cdots \circ (12\rangle(2)$.

60.3 Theorem (Howie)

The subsemigroup of T_n generated by the set of idempotents of rank $n-1$ consists of all transformations of rank $\leq n-1$.

Proof. We may assume that $n \geq 2$. Let $\alpha \in T_n$ be a transformation of rank $r \leq n-1$. Then $x, y, z, w \in \{1, \ldots, n\}$ exist such that $x\alpha = y\alpha = w$ and $z \notin \mathbf{r}\alpha$. Defining $\bar{\alpha} \in T_n$ by $x\bar{\alpha} = z$ and $u\bar{\alpha} = u\alpha$, for every $u \neq x$, we note that the rank of $\bar{\alpha}$ is $r+1$ and that $\alpha = \bar{\alpha} \circ (zw\rangle(w)$. We may therefore assume that the rank of α is $n-1$. In this case,

$$\alpha = (1 \cdots pi_{11}\rangle(i_{11} \cdots i_{1k_1})(i_{21} \cdots i_{2k_2}) \cdots (i_{m1} \cdots i_{mk_m}),$$

which may be expressed as

$$\alpha = (1 \cdots pi_{11}\rangle(i_{11} \cdots i_{1k_1}) \circ (1i_{21}\rangle(i_{21} \cdots i_{2k_2}) \circ \cdots \circ (1i_{m1}\rangle(i_{m1} \cdots i_{mk_m}).$$

So without loss of generality, we may assume that $\alpha = (1 \ldots pi_1\rangle(i_1 \ldots i_k)$, where $p \geq 1$ and $k \geq 1$. If $p = k = 1$, then α is an idempotent. Otherwise, the result follows from (1) of 60.2 (if $p \geq 2$ and $k = 1$), from (2) of 60.2 (if $p = 1$ and $k \geq 2$), or from (3) of 60.2 (if $p \geq 2$ and $k \geq 2$). □

§61 Comments

The time is ripe for studies of commutativity and centralizers in PT_n that parallel those presented in Chapters 3, 4, and 5. Some results have been obtained for full transformation semigroups T_n — there is the recent (1993) paper by A. Ehrenfeucht, T. Harju, and G. Rozenberg [1], where permutable transitive subsemigroups S and T of T_n are characterized as simply transitive groups of permutations that are centralizers of each other. That paper contains no pictures or references to digraphs. In contrast, the (1988) paper by Higgins [2], where those $\beta \in T_n$ that solve the equation $\alpha \circ \beta = \beta \circ \gamma$ $(\alpha, \gamma \in T_n)$ are constructed, and where those members $\alpha \in T_n$ whose centralizer $C(\alpha) = \{\alpha^m \mid m \geq 0\}$ are characterized, contains several pictures and references to functional digraphs (§56).

These diverse approaches (geometric versus algebraic) are typical. In this chapter, for instance, Theorem 58.8, which unifies the characterizations of commutativity in the hierarchy $S_n \subset C_n \subset PT_n$ — condition (ii) suffices for S_n, while (i) and (ii) suffice for C_n — may be viewed as a result obtainable by purely geometric methods. On the other hand, Lemma 60.2 (discovered in 1995 by Konieczny) is purely algebraic, requiring only multiplication, and led Konieczny and Lipscomb [2] to our proof of Howie's Theorem 60.3.

Earlier proofs of Howie's Theorem, namely Howie [2] (1965) and Higgins [2] (1992), are devoid of path notation. So one should ask, *Is path notation really necessary to study* PT_n? While it is a fact that path notation led to Theorem 58.8 and its spin-off Proposition 59.1, these results alone seem insufficient to justify the amount of effort required to develop such a notation. Currently, the state of path notation in PT_n is akin to its state in C_n prior to finding the alternating semigroups A_n^c — A_n^c grew out of practice with, and an understanding of, the possible path structures that could result from products of semitranspositions. This purely algebraic view (understanding the multiplication) was amplified when it proved to be the key to classifying the S_n-normal semigroups. In particular, the crown in that effort is the seemingly digraph uninspired Lemma 28.2, whose statement is merely a study in the path structures of products of conjugates, and whose proof (like Lemma 60.2) is trivial path notation multiplication.

The "foldings" $z \to z^m$ (Figure 58.4) that motivated the definition of a morphism of cells are common in topology, especially in the study of covering spaces (K. Jänich [1, Chapter IX]). For background on (the more general) morphisms of digraphs, see the text by Bondy and Murty [1].

CHAPTER 13

Centralizers, Conjugacy, Reconstruction

Beginning with centralizers in C_n, we reflect on the study in Chapters 3, 4, and 5, restating Lallement's question and proposing that the results on C_n be extended to PT_n (§62). We also return to conjugacy, considering Saito's Theorem [3] on the conjugacy of $\alpha \circ \beta$ and $\beta \circ \alpha$, counting conjugacy classes in C_n (§63), and introducing "conjugacy" in PT_n (§64). For the S_n-normal theory as it applies to PT_n, we propose that the classification presented in Chapter 7 be extended (§65).

Turning to reconstruction, we begin with an example that motivates the (semigroup) Reconstruction Conjecture (RC) (§66). The RC for semigroups concerns extensions of one Brandt semigroup by another and is equivalent to the classical RC for simple graphs. To show the equivalence, we use a category isomorphism (§67). The semigroup RC is then stated in §68.

§62 Centralizers

For a chart $\alpha = \gamma\eta \in C_n$ with permutation part γ and nilpotent part η, the centralizer $C(\alpha)$ relative to C_n is isomorphic to $C(\gamma) \times C(\eta)$. Expressing $\gamma = (\gamma_{11} \cdots \gamma_{1k_1}) \cdots (\gamma_{m1} \cdots \gamma_{mk_m})$ as a join of "regular charts" $\gamma_{i1} \cdots \gamma_{ik_i}$ with each circuit γ_{ij} having length ℓ_i, we recall from Chapter 4 that $C(\gamma)$ may be imbedded in a direct product $W_1 \times \cdots \times W_m$ of wreath products $W_i = Z_{\ell_i}^z$ wr C_{k_i}. On the other hand, expressing $\eta = \eta_1 \cdots \eta_u$ in its join representation with ℓ the maximum among the lengths of the η_t, we recall from Chapter 5 that $W = S^\ell$ wr C_u has a congruence $\sim$ such that $C(\eta)$ may be imbedded in the quotient $\mathcal{W} = W/\sim$. (In the wreath product, $S^\ell = \{\, \mu^k \varepsilon_j \,|\, 0 \leq j \leq \ell - 1;\ k \geq 0\}$ where $\mu = (0, 1, \ldots, \ell - 1]$ and $\varepsilon_j = (0) \cdots (j)$ elements in $C_{\{0,1,\ldots,\ell-1\}}$.)

Putting these structures together, we have an imbedding of $C(\alpha)$ into a product $W_1 \times \cdots \times W_m \times \mathcal{W}$ of wreath products W_i and the quotient $\mathcal{W}$:

$$C(\alpha) \;\simeq\; C(\gamma) \times C(\eta) \longrightarrow W_1 \times \cdots \times W_m \times \mathcal{W}.$$

This state of affairs brings us to Lallement's question [2] previously mentioned in §23.

62.1 Question

In view of the nature of the congruence $\sim$, *is it possible to obtain* $\mathcal{W}$ *as a wreath product of a transformation semigroup* X *and a quotient of a transformation semigroup* Y? *If the answer is yes, then we would have* $C(\alpha)$ *imbedded in simply a product of wreath products.*

The time is ripe to study centralizers in PT_n. For with the characterization 58.8 of commuting members of PT_n and the techniques developed for the C_n case (Chapters 3, 4, and 5), it may now be possible to find a construct that unifies the C_n and PT_n centralizer theories.

62.2 Problem

Extend the wreath products used in the C_n *centralizer case to wreath products that accommodate centralizers in* PT_n. *Use* 58.8 *to find a formula for computing the order* $|C(\alpha)|$ *of thc centralizer* $C(\alpha)$ *in* PT_n.

In computing orders $|C(\alpha)|$ when $\alpha = \varepsilon$ is an idempotent in PT_n, recall the solution 59.1. For full transformations, the bottleneck is in counting the mappings from cells onto subcells.

§63 Conjugacy in C_n

Recall first that a chart $\alpha \in C_n$ is *conjugate* to a chart $\beta \in C_n$ if a permutation $\gamma \in S_n$ exists such that $\gamma^{-1} \circ \alpha \circ \gamma = \beta$. And recall second that charts are conjugate if and only if they have the same path structure, showing that conjugacy in C_n is a generalization of conjugacy in S_n. This generalization, however, does not run parallel to the S_n case. For instance, if α and β are permutations, then we always have $\alpha \circ \beta$ conjugate to $\beta \circ \alpha$ because $\alpha^{-1} \circ (\alpha \circ \beta) \circ \alpha = \beta \circ \alpha$. But for some charts α and β, the product $\alpha \circ \beta$ may not be conjugate to $\beta \circ \alpha$, as $\alpha = (1](2345]$ and $\beta = (2](1)(3)(4)(5)$ demonstrate — $\alpha \circ \beta = (1](2345]$ is not conjugate to $(1](2](345] = \beta \circ \alpha$.

In 1991, T. Saito [3] found a rather nice characterization for two charts to have their two products conjugate. He was able to detail a matching between the paths in the path decomposition of $\alpha \circ \beta$ with those in the decomposition of $\beta \circ \alpha$ that must occur whenever the two products are conjugate. To explain his solution, we first need some notation. For $\alpha \in C_n$, we (generically) denote the left and right endpoints of any maximal proper path in α via

$$\alpha = \cdots (i^\alpha \cdots t^\alpha] \cdots .$$

If $\alpha = (123](45]$, for example, then each $i^\alpha \in \{1, 4\} = I_\alpha$ is an *initial point*, and each $t^\alpha \in \{3, 5\} = T_\alpha$ is a *terminal point*. (Saito [3] denotes I_α and T_α, respectively, as Y_α and X_α.) Continuing, we also represent points in $\mathbf{d}\alpha$ and $\mathbf{r}\alpha$, respectively, by d^α and r^α.

It is then easy to show that the form of the path decompositions of the two possible orders of products of $\alpha, \beta \in C_n$ are

$$\begin{aligned} \alpha\circ\beta &= \cdots(i^\beta\cdots d^\alpha]\cdots(i^\beta\cdots t^\alpha]\cdots(r^\beta\cdots d^\alpha]\cdots(r^\beta\cdots t^\alpha]\cdots \\ \beta\circ\alpha &= \cdots(r^\alpha\cdots t^\beta]\cdots(r^\alpha\cdots d^\beta]\cdots(i^\alpha\cdots t^\beta]\cdots(i^\alpha\cdots d^\beta]\cdots . \end{aligned} \tag{63.1}$$

If a path having form $(i^\beta\cdots d^\alpha]$ appears in $\alpha\circ\beta$, then, applying α to each point in $\{i^\beta,\dots,d^\alpha\}$, we see that α maps $(i^\beta\cdots d^\alpha]$ onto a path in $\beta\circ\alpha$ having form $(r^\alpha\cdots t^\beta]$. And these two paths must have the same length.

So in the configuration in (63.1), we may use α to map "down." Similarly, using β to map "up," we have another match (vertically) between the path forms on the far right in (63.1). Such "outside matchings" hold for any pair of products, independent of whether the products are conjugate.

Saito's insight is that the "cross-matching" of the forms $(i^\beta\cdots t^\alpha]$ in $\alpha\circ\beta$ with those $(i^\alpha\cdots t^\beta]$ in $\beta\circ\alpha$ is both necessary and sufficient for $\alpha\circ\beta$ to be conjugate to $\beta\circ\alpha$.

63.2 Theorem (Saito)

Let $\alpha,\beta\in C_n$. Then $\alpha\circ\beta$ is conjugate to $\beta\circ\alpha$ if and only if $\alpha\circ\beta$ having a proper k-path $(i^\beta\cdots t^\alpha]$ is equivalent to $\beta\circ\alpha$ having a proper k-path $(i^\alpha\cdots t^\beta]$.

To illustrate 63.2, first consider $\alpha = (1](23456789)$ and $\beta = (2](13456789)$, and then calculate that $\alpha\circ\beta$ is conjugate to $\beta\circ\alpha$, i.e.,

$$\begin{aligned} \alpha\circ\beta &= (\overset{i^\beta}{2}468\overset{t^\alpha}{1}](\overset{r^\beta}{3}57\overset{d^\alpha}{9}] \quad\text{and} \\ \beta\circ\alpha &= (\underset{r^\alpha}{3}57\underset{d^\beta}{9}](\underset{i^\alpha}{1}468\underset{t^\beta}{2}]. \end{aligned}$$

And, if $\alpha = (1](23456789)$ and $\beta = (2](13)(456789)$, then $\alpha\circ\beta$ is not conjugate to $\beta\circ\alpha$, i.e.,

$$\begin{aligned} \alpha\circ\beta &= (\overset{i^\beta}{2}\,\overset{t^\alpha}{1}](\overset{r^\beta}{3}57\overset{d^\alpha}{9}](468) \quad\text{and} \\ \beta\circ\alpha &= (\underset{r^\alpha=d^\beta}{3}](\underset{i^\alpha}{1}468\underset{t^\beta}{2}](579). \end{aligned}$$

The down side of Saito's theorem is that, in practice, we must do at least some multiplications prior to its application. In other words, it would also be nice to have other characterizations of conjugacy for two products of α and β. In particular, if α has path decomposition $(1]\gamma$, with permutation part γ, and β has decomposition $(2]\delta$, with permutation part δ, then when is $\alpha\circ\beta$ conjugate to $\beta\circ\alpha$? A nice solution (if possible) would merely require inspection of the path decomposition of α relative to that of β, with no multiplications required.

63.3 Problem

Find wide classes S of charts in C_n such that whenever $\alpha \in S$, then we may determine whether $\alpha \circ \beta$ is conjugate to $\beta \circ \alpha$ by only inspecting the path structure of α relative to β.

Along this line of thought, we have the following result, a result that may be viewed as a corollary to Saito's Theorem.

63.4 Corollary

If $\alpha \in S_n$, then for any $\beta \in C_n$, the product $\alpha \circ \beta$ is conjugate to $\beta \circ \alpha$.

Proof. Note that if $\alpha \in S_n$, then we have the classical argument $\alpha^{-1} \circ (\alpha \circ \beta) \circ \alpha = \beta \circ \alpha$. But we may also use 63.2: For $\beta \in C_n$ and any $k \geq 1$, the form $(i^\beta \cdots t^\alpha]$ cannot exist in the path decomposition of $\alpha \circ \beta$ because there are no terminal points t^α of α. And since α also has no initial points i^α, the form $(i^\alpha \cdots t^\beta]$ cannot exist in the path decomposition of $\beta \circ \alpha$. Thus, by 63.2, the two products must be conjugate. □

If α and β are conjugate charts, then $\langle \alpha \rangle$ is isomorphic to $\langle \beta \rangle$. However, $\langle \alpha \rangle$ may be isomorphic to $\langle \beta \rangle$ with α not conjugate to β, as $\alpha = (12](34](56)$ and $\beta = (12](34)(56)$ illustrate.

The conjugacy tie between permutation groups and inverse semigroups also appears in a combinatorial form. For example, our next theorem (counting conjugacy classes of C_n) should be compared to its classical group counterpart (see Problem 1 on page 25 of Lovász [1]).

63.5 Theorem (Counting Conjugacy Classes)

The number of conjugacy classes in C_n is $\sum_{k=0}^{n} \pi(k)\pi(n-k)$, where $\pi(k)$ is the number of partitions of the integer k, and $\pi(0) = 1$.

Proof. A conjugacy class may be described by the cardinalities of circuits and proper paths: Given n, for each $k \leq n$ partition n into two parts $n = k + (n-k)$. Then think of k as the number of "circuit vertices," and $n-k$ the number of "proper path vertices." If we let $\pi(k)$ and $\pi(n-k)$ denote the number of partitions of the integers k and $n-k$, respectively, we see that the number of conjugacy classes must be as stated above. □

To illustrate 63.5, let us calculate the number of conjugacy classes in C_3.

Using the formula, we have $2[\pi(0)\pi(3) + \pi(1)\pi(2)] = 10$. In fact, there are three classes $(-](-](-]$, $(--](-]$, and $(---]$ that have all paths as proper paths; there are another three classes $(-)(-)(-)$, $(--)(-)$, and $(---)$ that have all paths as circuits; and there are yet another four classes $(--)(-]$, $(-)(-)(-]$, $(-)(--]$, and $(-)(-](-]$ of mixed variety.

§64 Conjugacy in PT_n

If we define $\alpha, \beta \in PT_n$ to be conjugate when there exists a permutation $\gamma \in S_n$ such that $\gamma^{-1} \circ \alpha \circ \gamma = \beta$, then it is *not* true that α and β are conjugate if and only if α and β have the same path structure. For instance, in PT_5 consider that $\alpha = (12\rangle(32\rangle(245)$ and $\beta = (12\rangle(34\rangle(245)$ have the same path structure (the proper paths and circuits can be matched so that lengths are preserved) but are not conjugate. Nevertheless, it is easy to see that "same path structure" is an equivalence relation on PT_n. It is also obvious that conjugate partial transformations necessarily have the same path structure.

64.1 Problem

Find a (significant) application in PT_n for the relation "same path structure" where the conjugacy relation is deficient. In other words, show that the relation "same path structure" is worthy of further study.

§65 S_n-normal Semigroups in PT_n

As mentioned in §35, for finite X, Symons [1] (in 1976) classified the $\mathcal{G}_X$-normal semigroups of full transformations. In the more general (partial transformation) PT_n case, a similar "to appear" result was announced in 1980 by Sullivan [3, Ref. 8].

65.1 Problem

Classify the S_n-normal subsemigroups of PT_n. Such a classification would extend the classification for both the C_n (Chapter 7) *and the full transformation cases* (Symons [1]).

§66 Graph Reconstruction

Let $N^{(2)}$ be the collection of doubleton subsets of $N = \{1, 2 \ldots, n\}$ $(n \geq 3)$, let $E \subset N^{(2)}$, and suppose $V \subset N$ is a nonempty set that *spans* E, i.e.,

$\cup\{ e \mid e \in E \} \subset V$. Then each ordered pair $G = (V, E)$ is a *simple graph* with *vertex set* V and the *edge set* E. For morphisms of simple graphs, $G = (V, E)$ is *isomorphic* to $H = (V', E')$ when a bijection $\phi : V \to V'$ exists such that $e\phi \in E'$ if and only if $e \in E$.

For each vertex $i \in V$ of a simple graph $G = (V, E)$, the graph $G_i = G - i$ is obtained from G by deleting the vertex i and its incident edges, i.e., G_i is the *vertex-deleted subgraph* of G that has vertex set $V - \{i\}$ and edge set $E - \{ e \in E \mid i \in e \}$. One of the most well-known unsolved problems of graph theory asks whether a graph can be reconstructed up to isomorphism if we know all of its vertex-deleted subgraphs up to isomorphism. This reconstruction conjecture may be stated as follows.

Graph Reconstruction Conjecture. *If $G = (V, E)$ and $H = (V, E')$ are simple graphs with at least 3 vertices, and if G_i is isomorphic to H_i for each vertex $i \in V$, then G is isomorphic to H.*

To motivate an equivalent conjecture for semigroups, let us begin by orienting the edges of a simple graph, i.e., let us consider a special kind of *digraph*, namely, a digraph $D = (V, A)$ where $V \subset N$, where $A \subset N \times N$ is a set of *arcs* such that $(i, i) \notin A$, for every $i \in N$, and $(i, j) \in A$ implies $(j, i) \notin A$, and where $(i, j) \in A$ implies $\{i, j\} \subset V$.

One such digraph D is pictured in Figure 66.1, where we also see a corresponding inverse semigroup D'. Note that the semigroup D' contains the Brandt semigroup $I_1 \subset C_3$ of all charts of rank ≤ 1 as an ideal, and that the Rees quotient D'/I_1 is also a Brandt semigroup. It follows that D' is an extension of one Brandt semigroup (with trivial group) by another (also with trivial group). In general, each such (simple) digraph D, with vertices $1, 2, \ldots, n \geq 3$, yields an inverse subsemigroup D' of C_n comprised of all charts of rank ≤ 1 and those charts of rank 2 that map the vertices of an arc in D onto the vertices of an arc in D while preserving orientation. Moreover, every such D' is an extension of one Brandt semigroup I_1 by another D'/I_1.

Each digraph D, with vertex set $\{1, 2, \ldots, n\}$ $(n \geq 3)$, also has a *deck*, which is a multiset $\mathcal{D} = \{D_1, D_2, \ldots, D_n\}$ of isomorphic copies of the vertex-deleted subgraphs of D, i.e., D_i is isomorphic to the maximal subgraph of D having vertex set $\{1, 2, \ldots, n\} - \{i\}$. The graph D_i is obtained from D by removing the vertex i and all incident diedges. The members of a deck are sometimes called *cards*. For example, the three cards (in the deck) of the digraph pictured in Figure 66.1 may be constructed by removing the labels 1, 2, and 3 on the digraphs in Figure 66.2.

Similarly, the semigroup D' in Figure 66.1 has a deck, which is a multiset $\mathcal{D}' = \{D'_1, D'_2, D'_3\}$ of semigroups. In this case, each D'_i is isomorphic to $\{ \alpha \in D' \mid i \notin (\mathbf{d}\alpha \cup \mathbf{r}\alpha) \}$. The cards D'_i (in the deck) of D' may also be constructed with the help of Figure 66.2: For D'_1, start on the left and replace the label 2 with x and the label 3 with y; for D'_2, move to the middle

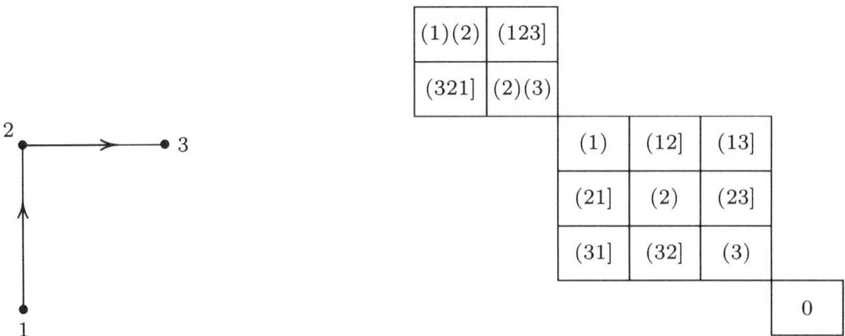

Figure 66.1. A digraph D and corresponding semigroup D'.

and replace 1 with u and 3 with v; and for D'_3, move to the right and replace 1 with p and 2 with q. In other words, cards in the deck of D' exhibit no connection via any set of labels. (The labels in Figure 66.2 show how to "paste" the cards to reconstruct the D' in Figure 66.1.)

It turns out that nonisomorphic digraphs may have the same deck. For instance, the digraphs in Figures 66.1 and 66.3 are nonisomorphic but nevertheless have the same deck. A similar statement holds for the nonisomorphic semigroups in Figures 66.1 and 66.3, even though both are extensions of I_1 by the same semigroup.

For arbitrary extensions of one Brandt semigroup by another, Lallement and Petrich [1] found a fundamental connection to certain collections of sets. For reconstruction of digraphs, Stockmeyer [1]—[6] and Kocay [1] found infinite classes of "nonreconstructible" digraph-pairs. But neither the extension characterization nor the digraph discoveries has produced any classification theorems concerning extensions of one Brandt semigroup by another.

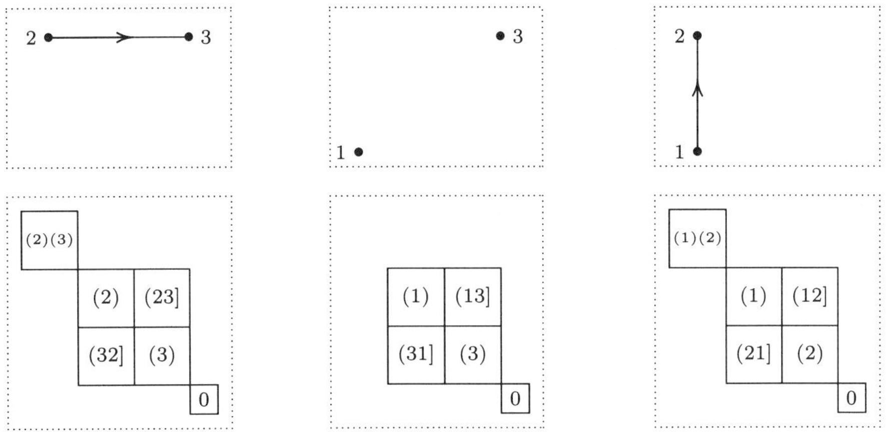

Figure 66.2 Digraph and corresponding semigroup decks.

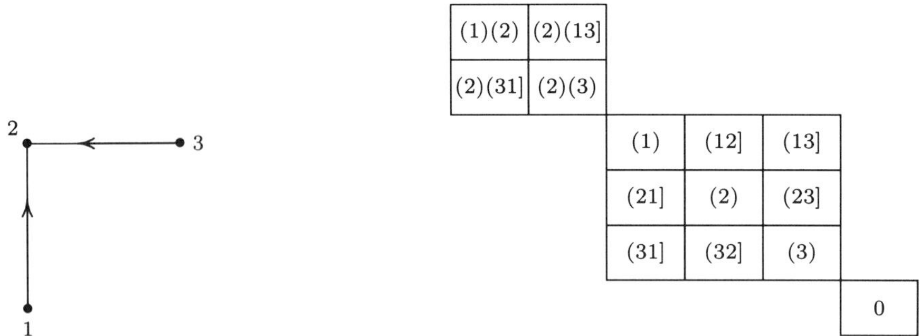

Figure 66.3 Another digraph and corresponding semigroup.

66.4 Problem

Find simple conditions on the extensions of one Brandt semigroup, with trivial group, by another, also with trivial group, that explain the counterexamples to the digraph reconstruction conjecture.

§67 Isomorphic Categories of Graphs and Semigroup Extensions

We define a functor $* : \mathcal{G} \to \mathcal{E}$ from a category $\mathcal{G}$ of simple graphs to a category of semigroups $\mathcal{E}$: With each simple graph $G = (V, E)$, we associate an extension $G_* = G_*(S, T)$ of one Brandt semigroup S, with 1-group (trivial group), by another T, with 2-group. In particular,

$$S = \{\, \alpha \in C_n \mid \text{ rank of } \alpha \text{ is one; } \mathbf{d}\alpha, \mathbf{r}\alpha \subset V \,\} \cup \{0\}$$

has chart composition for its multiplication; and

$$T = \{\, \alpha \in C_n \mid \text{ rank of } \alpha \text{ is two; } \mathbf{d}\alpha, \mathbf{r}\alpha \in E \,\} \cup \{0\}$$

has chart composition for its multiplication when the product is a member of the set T, and has all other products defined as zero. The extension $G_* = G_*(S, T)$ of S by T has underlying set $S \cup T$ and its multiplication is that of C_n, i.e., G_* is a subsemigroup of C_n. Note that G_*, S, and T are induced, respectively, from G, V, and E.

The domain of the functor $* : \mathcal{G} \to \mathcal{E}$ is the category $\mathcal{G}$ with *objects*: simple graphs $G = G(V, E)$; and *morphisms*: graph morphisms that are "into-isomorphisms." For instance, if $f : G \to H$ is a morphism in $\mathcal{G}$, then it may be true that $f(G) \neq H$. But f is always an isomorphism from its domain G onto its range $f(G) \subset H$.

The category $\mathcal{E}$ has *objects*: inverse semigroups G_*; and *morphisms*: semigroup morphisms that are rank-restricted into-isomorphisms. A morphism

ϕ is *rank-restricted* when $\text{rank}(\alpha) = \text{rank}(\phi(\alpha))$ for all $\alpha \in \mathbf{d}\phi$. A rank-restricted morphism $f_* : G_* \to H_*$ is a morphism in $\mathcal{E}$ when f_* is an isomorphism from its domain G_* onto its range $f_*(G_*) \subset H_*$. To illustrate the kinds of isomorphisms that are not in $\mathcal{E}$, let $G_* = \{ (1], (1), (2), (12], (21] \}$ and

$$H_* = \{ \alpha \in C_4 \mid \text{rank } \alpha \le 1 \} \ \cup \ \{ (1)(2),(12), (3)(4), (34), (13](24], \\ (14](23], (31](42], (41](32] \}$$

be subsemigroups of C_4, and let $\phi : G_* \to H_*$ be given by $(1] \mapsto (1]$, $(1) \mapsto (1)(2)$, $(2) \mapsto (3)(4)$, $(12] \mapsto (13](24]$, and $(21] \mapsto (31](42]$.

The functor $* : \mathcal{G} \to \mathcal{E}$ is given by $G \mapsto G_*$, and $f \mapsto f_*$ where (with operators on the left)

$$f_*(\alpha) = f \circ \alpha \circ f^{-1} \in H_* \text{ for each } \alpha \in G_* :$$

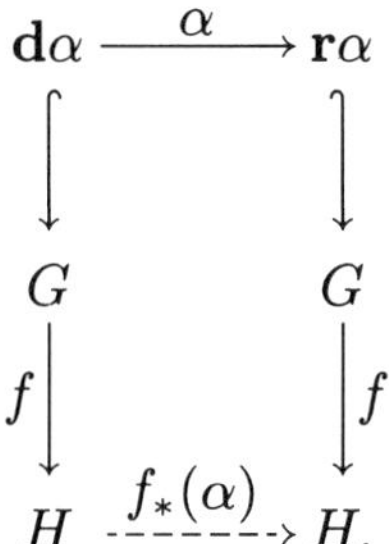

To see that f_* is indeed a morphism in $\mathcal{E}$, we first note that f_* is a semigroup homomorphism (standard argument). To see that f_* is injective, let $\alpha, \beta \in G_*$ and suppose $\alpha \ne \beta$. Then one of these two charts is defined at some i where the other is undefined, in which case $f_*(\alpha) \ne f_*(\beta)$, or $\alpha(i) \ne \beta(i)$ for some i, and we let $f^{-1}(j) = i$. Since f is injective,

$$f_*(\alpha)(j) = f \circ \alpha(f^{-1}(j)) \ne f \circ \beta(f^{-1}(j)) = f_*(\beta)(j),$$

showing $f_*(\alpha) \ne f_*(\beta)$. And finally, to see that f_* is rank-restricted, note that since f is injective and since

$$\mathbf{d}f_*(\alpha) = \mathbf{d}(f \circ \alpha \circ f^{-1}) = f(\mathbf{d}\alpha),$$

we have $|\mathbf{d}f_*(\alpha)| = |\mathbf{d}\alpha|$. It follows that $* : \mathcal{G} \to \mathcal{E}$ is a functor.

To define the $*$-inverse functor $\natural$, we shall denote rank-1 (rank-2) idempotents that fix i (i and j) by ε_i (ε_{ij}). With this notation, we let $\natural : \mathcal{E} \to \mathcal{G}$ be given by $G_* \mapsto G$ and $h \mapsto h_\natural$ where, for $h : G_* \to H_*$, we let $h_\natural$ be given by

$$h(\varepsilon_i) = \varepsilon_{h_\natural(i)} \text{ for each vertex } i \text{ of } G. \tag{67.1}$$

To see that $h_\natural : G \to H$ is indeed a morphism in $\mathcal{G}$, we first observe that it is well-defined since h is rank-restricted. That it is injective follows because h is injective. To see that it is a graph into-isomorphism, first observe that for vertices i and j of G,

$$h(\varepsilon_{ij}) = \varepsilon_{h_\natural(i)h_\natural(j)}, \tag{67.2}$$

where (67.2) follows since

$$h(\varepsilon_i \circ \varepsilon_{ij}) = h(\varepsilon_i) \circ h(\varepsilon_{ij}) = \varepsilon_{h_\natural(i)} \circ h(\varepsilon_{ij}) \Rightarrow h_\natural(i) \in \mathbf{d}h(\varepsilon_{ij}).$$

Similarly, $h_\natural(j) \in \mathbf{d}h(\varepsilon_{ij})$. We now apply (67.2) and the reasoning indicated in the following "diagram," where horizontal left-right arrows mean *if and only if*

$$\begin{array}{ccc} \{i,j\} \in E(G) & \longleftrightarrow & \varepsilon_{ij} \in G_* \\ h_\natural \big\downarrow & & \big\downarrow h \\ \{h_\natural(i), h_\natural(j)\} \in E(H) & \longleftrightarrow & \varepsilon_{h_\natural(i)h_\natural(j)} \in H_*. \end{array}$$

67.3 Proposition **(Representation of h in terms of $h_\natural$.)**

Given $h : G_* \to H_*$ in $\mathcal{E}$, and $h_\natural : G \to H$, then $h(\alpha) = h_\natural \circ \alpha \circ h_\natural^{-1}$ for every $\alpha \in G_*$.

Proof. It suffices to observe that if $\alpha \in G_*$ and if α moves i to j, then $h(\alpha)$ moves $h_\natural(i)$ to $h_\natural(j)$. But this observation follows since $h(\varepsilon_i \circ \alpha \circ \varepsilon_j) = h(\varepsilon_i) \circ h(\alpha) \circ h(\varepsilon_j) = \varepsilon_{h_\natural(i)} \circ h(\alpha) \circ \varepsilon_{h_\natural(j)}$ and h preserves rank. $\square$

Again, a standard argument shows that $\natural : \mathcal{E} \to \mathcal{G}$ is a functor.

67.4 Theorem **(Categories $\mathcal{G}$ and $\mathcal{E}$ are Isomorphic.)**

The functors $*$ and $\natural$ are category isomorphisms, i.e., $*\circ\natural = 1_{\mathcal{E}}$ and $\natural\circ* = 1_{\mathcal{G}}$.

Proof. These functors are clearly mutually inverse on objects. So consider these functors on morphisms. For $h : G_* \to H_*$ in $\mathcal{E}$, $(h_\natural)_*(\alpha) = h_\natural \circ\alpha\circ h_\natural^{-1} = h(\alpha)$. Thus, $* \circ \natural = 1_{\mathcal{E}}$. For $f : G \to H$ in $\mathcal{G}$,

$$f_*(\varepsilon_i) = f \circ \varepsilon_i \circ f^{-1} = \varepsilon_{f(i)}$$

and from (67.1)

$$f_*(\varepsilon_i) = \varepsilon_{(f_*)_\natural(i)}.$$

Thus, for all vertices i of G, $f(i) = (f_*)_\natural(i)$, showing $(f_*)_\natural = f$. In other words, $\natural \circ * = 1_{\mathcal{G}}$. $\square$

In summary, both categories $\mathcal{G}$ and $\mathcal{E}$ were obtained by essentially choosing edge sets. While the idea of a graph is clearly obtained by such a process, it was Lallement and Petrich [1] who showed that extensions of one Brandt semigroup by another is also fundamentally related to choosing collections of sets whose pairwise intersections are at most singleton sets.

§68 Semigroup Reconstruction Conjecture

When there is no confusion, we shall drop the $*$-subscript on the objects in $\mathcal{E}$. In particular, we call G_i a *vertex-deleted subsemigroup of* $G \in \mathcal{E}$ whenever G_i is obtained by removing from G all charts that have i in either their domain or their range. We shall also say that $G_* \in \mathcal{E}$ has *vertex set* V when $G = G(V, E)$.

Semigroup Reconstruction Conjecture. *If each of* $G, H \in \mathcal{E}$ *has vertex set* $N = \{1, 2, \ldots, n\}$ $(n \geq 3)$, *and if* G_i *is isomorphic to* H_i *for every vertex* $i \in N$, *then* G *is isomorphic to* H.

By design, this conjecture may be transformed into the Graph Reconstruction Conjecture by replacing "$\mathcal{E}$" with "$\mathcal{G}$." And with the category isomorphism $* : \mathcal{G} \to \mathcal{E}$, we can straightforwardly see that a solution to either conjecture yields a solution to the other. As to progress up to 1977 on the graph version, Bondy and Hemminger [1] provide a survey. In the same volume of that journal, McKay [1] reports that computer programs have verified the reconstruction conjecture for graphs (and hence for semigroups) with at most 9 vertices. On the other hand, the orthodox approach to finding a counterexample is explained in a 1981 paper by Kimble, Schwenk, and Stockmeyer [1], where the idea of *pseudo-similarity* is fundamental. They state, "We suggest that any search for nonreconstructible graphs should begin at this size (20 vertices)." A 1988 result by Yang Yongzhi [1] shows that one only need to prove the conjecture true for graphs that are 2-connected.

68.1 Problem

Apply the known results on graph reconstruction to the corresponding extension problem in semigroups. In particular, since the reconstruction conjecture is known to be true for disconnected graphs and trees, we need a (nice) equivalence between connected graphs and extensions of one Brandt semigroup (with trivial group) by another (with 2*-group).*

In thinking about this problem it may be helpful to consider the fact that each path $v_1, e_1, v_2, e_2, \ldots, v_k, e_k, v_{k+1}$ in a simple graph gives rise to a product of the form $(v_1, v_2, v_3] \circ (v_2, v_3, v_4] \circ \cdots \circ (v_{k-1}, v_k, v_{k+1}]$.

§69 Comments

For background in graph theory, see either Bondy and Murty [1] or Harary [2]. For graph reconstruction, see either Bondy and Hemminger's survey [2] or Harary's paper [3]. Saito's Theorem 63.2 first appeared in Saito [3]. The category isomorphism in §67 first appeared in Lipscomb [5].

Appendix

We briefly review some of the basic concepts. The introductory section (§70) provides the first level of definitions. At the next level, we present the relevant aspects of abstract groups (§71) and permutation groups (§72). We then focus on centralizers in groups (§§73 – 74). We also consider certain aspects of semigroups (§§75 – 76). Green's relations are defined (§77) and free algebras are mentioned (§78). Our final section (§79) concerns categories.

§70 Objects and Elements

Given a nonempty set S, we frequently call a map (binary operation) $S \times S \to S$ a *multiplication* (on S) and then refer to its values as *products.* The product "ab" of elements a and b is just the image of $(a, b) \in S \times S$.

A multiplication $S \times S \to S$ is *associative* if it satisfies the associative law:

$$(ab)c = a(bc) \quad (a, b, c \in S).$$

That is, for each string abc of elements of S, the induced iterated products $(ab)c$ and $a(bc)$ are equal. The associative law implies the *general associative law*, which states that for each string $a_1 a_2 \cdots a_k$ of elements of S, all induced iterated products are equal.

For an associative multiplication, the product $A_1 A_2 \cdots A_k$ of subsets of S is given by

$$A_1 A_2 \cdots A_k = \{\, a_1, \ldots , a_k \,|\, a_i \in A_i \text{ for each } i, 1 \leq i \leq k \,\}.$$

When $k = 2$ and $A_1 = \{x\}$ is a singleton set, we denote $\{x\}$ as x and thereby obtain $xA = \{x\}A = \{\, xa \,|\, a \in A\}$.

A nonempty set S together with an associative multiplication $S \times S \to S$ is a *semigroup.* As is customary, both a semigroup and its underlying set are tagged with the same letter. This double entendre provides a flexibility in expressing definitions: If S is a semigroup and T is a nonempty subset of S, then T is a *subsemigroup* of S if $a, b \in T$ implies $ab \in T$.

An element 1 of a semigroup S is an *identity* if $a1 = 1a = a$ for every $a \in S$. Each semigroup clearly contains at most one identity, and those that contain identities are called *monoids.* A *submonoid T of a monoid M*

is any subsemigroup of M that contains the identity of M. In contrast, a *submonoid* T *of a semigroup* S is a subsemigroup of S that is also a monoid.

If 1 is the identity of the monoid M, then $u \in M$ is a *unit* if there exists $v \in M$ such that $uv = vu = 1$. If such a v exists, it certainly depends on u and it is clearly unique. To express this unique dependence, we usually write $v = u^{-1}$ and call u^{-1} the *inverse* of u. A monoid G in which every $g \in G$ is a unit is called a *group*. As for subgroups of groups, if G is a group and H is a subset of G, then H is a *subgroup of* G if H is a submonoid of the monoid G such that $u \in H$ implies $u^{-1} \in H$. In contrast, a submonoid H of a semigroup S is a *subgroup of* S if H is a also a group. We note in passing that each monoid M contains a subgroup, namely, its *group of units* $G = \{\, u \in M \mid u \text{ is a unit } \}$.

§71 Groups and Group Morphisms

Let G and H be groups and suppose $\phi : G \to H$ is a function that *preserves multiplication*, i.e.,

$$ab\phi = a\phi\, b\phi \quad (a, b \in G).$$

Then ϕ is called a *morphism*, a *homomorphism*, or a *representation*. An injective morphism $G \to H$ is a *monomorphism*, a *faithful representation*, or an *imbedding* (of G into H); a surjective morphism is an *epimorphism*; and a bijective morphism is an *isomorphism*.

Morphisms are determined by *normal subgroups* (*self-conjugate subgroups*) — a subgroup K of G is normal (or self-conjugate) if $g^{-1}Kg = K$ for every $g \in G$. Given a normal subgroup K of G, the *factor group* G/K of *left cosets* gK $(g \in G)$ has multiplication $(g_1K)(g_2K) = (g_1g_2)K$. And the (natural) epimorphism $\natural : G \to G/K$ maps $g \mapsto gK$. An element $g \in G$ is called a *representative* of gK; and $gK = hK$ if and only if $gh^{-1} \in K$. When G is finite, the number of elements $|G/K|$ in G/K is $|G|/|K|$ and is called the *index* of K in G.

Conversely, normal subgroups are determined by morphisms.

71.1 Fundamental Homomorphism Theorem

Let G and H be groups, with 1_H denoting the identity of H, and let $\phi : G \to H$ be a homomorphism. Then,

$$K = \ker \phi \;= \{\, g \in G \mid g\phi = 1_H \,\}$$

is a normal subgroup of G; the map $\natural : G \to G/K$ given by $g \mapsto gK$ is an epimorphism; the map $\alpha : G/K \to G\phi \subset H$ given by $gK \mapsto g\phi$ is an isomorphism; and $\natural \circ \alpha = \phi$.

The subgroup "$\ker\phi$" given in 71.1 is often called the *kernel of* ϕ. With each kernel being a normal subgroup and with each normal subgroup being a kernel, we may deduce from 71.1 that the study of group morphisms is essentially a study of normal subgroups.

Turning to generators and generating sets, we say that a subset A of a group G *generates* G if every $g \in G$ is a product of members of $A \cup A^{-1}$ (where $A^{-1} = \{ a^{-1} \mid a \in A\}$). In this case, each $a \in A$ is called a *generator* of G. More generally, if A is any subset of G, then the *subgroup H generated by A* is the intersection of all subgroups of G that contain A; and each $a \in A$ is called a *generator* of H. In the case where $A = \{a\}$ is a singleton subset of G, we denote the subgroup of G generated by A as $\langle a \rangle$. The subgroup $\langle a \rangle$ consists of all integral powers a^n of a and it is called the *cyclic subgroup generated by a*. Whenever $G = \langle a \rangle$, we say that G is a *cyclic group*. For example, Z_n denotes the cyclic group $\langle 1 \rangle$ whose underlying set is $\{0, 1, \ldots, n-1\}$ and whose multiplication is addition modulo n. These cyclic groups Z_n, for $n = 1, 2, \ldots$, serve as models for all finite cyclic groups — if G is cyclic of order n, then G is isomorphic to Z_n.

Elements a and b of a group G are *conjugate* when there exists a $g \in G$ such that $g^{-1}ag = b$. It is easy to see that this relation of conjugacy is an equivalence relation. In addition, each (self-conjugacy) equation $b^{-1}ab = a$ produces an *element b that commutes with a*, i.e., $ab = ba$. The set of all elements in a group G that commute with $a \in G$ is a subgroup of G. It is denoted $C_G(a)$, i.e.,

$$C_G(a) = \{ b \in G \mid ab = ba \},$$

and it is called the *centralizer of a* (in G). When the underlying group G is clear from the context of the discussion, we suppress the G and simply denote the centralizer of a in G as $C(a)$.

§72 Permutation Groups

For any set X, a bijective map $X \to X$ is called a *permutation* (of X). The set S_X of all permutations of X together with function composition $S_X \times S_X \to S_X$ is a group, the *symmetric group S_X on X*. Subgroups of symmetric groups are called *permutation groups*; and permutation groups were the precursors of abstract group theory. The study of abstract groups is, however, in the final analysis, a study of permutation groups. This fact is a result of the following well-know theorem.

72.1 Theorem (Cayley)

Let G be any group. Then G is isomorphic to a permutation group.

Since the standard proof of Cayley's Theorem imbeds G into S_G, it follows that any finite group that has n elements may by imbedded in $S_{\{1,2,\dots,n\}}$, which we shall denote as S_n.

If $X = \{x_1, \dots, x_n\}$ contains n elements, then the *standard notation* for a permutation $\alpha \in S_X$ is

$$\alpha = \begin{pmatrix} x_1 & x_2 & x_3 & \cdot & \cdot & \cdot & x_n \\ x_1\alpha & x_2\alpha & x_3\alpha & \cdot & \cdot & \cdot & x_n\alpha \end{pmatrix}.$$

A more flexible and useful notation is *cycle notation*: Let $\alpha \in S_n$, where $N = \{1, \dots, n\}$. We say that α *fixes* $i \in N$ if $i\alpha = i$; otherwise, we say that α *moves* i. In particular, let $i_1, \dots, i_m$ be distinct integers in N and suppose $M = \{i_1, \dots, i_m\}$. If $\alpha \in S_n$ fixes each of the k members of $N - M = \{j_1, \dots, j_k\}$ according to

$$\alpha = \begin{pmatrix} i_1 & i_2 & i_3 & \cdot & \cdot & \cdot & i_m & j_1 & j_2 & \cdot & \cdot & \cdot j_k \\ i_2 & i_3 & i_4 & \cdot & \cdot & \cdot & i_1 & j_1 & j_2 & \cdot & \cdot & \cdot j_k \end{pmatrix},$$

then α is a *cycle of length* m or, more simply, an m*-cycle*, and the (cycle) notation for α is $(i_1 i_2 \cdots i_m)$. For example, $(123) \in S_5$ denotes the 3-cycle that fixes 4 and 5 while moving 1 to 2 and 2 to 3 and 3 to 1. Every 1-cycle is the identity permutation of N because it fixes every element of N.

Two *permutations are disjoint* if every point moved by one is fixed by the other. Disjoint permutations, and hence disjoint cycles, clearly commute. The useful feature of disjoint cycles is that every permutation is a product of cycles that are pairwise disjoint.

72.2 Theorem

Let S_n be the symmetric group on $N = \{1, 2, \dots, n\}$, and let $\alpha \in S_n$. Then

$$\alpha = \alpha_1 \circ \alpha_2 \circ \cdots \circ \alpha_m$$

where each α_i, $1 \le i \le m$, is a ℓ_i-cycle with $\sum_{i=1}^m \ell_i = n$, and, for distinct indices i and j, α_i and α_j are disjoint. Moreover, except for the order in which the α_i's are written, this factorization is unique.

The unique factorization guaranteed by 72.2 is called the *cycle decomposition* of the permutation $\alpha \in S_n$. More precisely, we may write

$$\alpha = (a_{11}a_{12}\cdots a_{1\ell_1}) \circ (a_{21}a_{22}\cdots a_{2\ell_2}) \circ \cdots \circ (a_{m1}a_{m2}\cdots a_{m\ell_m}), \tag{72.3}$$

where $\sum_{i=1}^m \ell_i = n$ and $a_{ij} \in N$; and, when $\ell_i, \ell_k > 1$, $a_{ij} = a_{kl}$ implies $i = k$ and $j = l$ with

$$a_{ij}\alpha = a_{ij+1} \quad (j < \ell_i), \qquad a_{i\ell_i}\alpha = a_{i1}.$$

By 72.2 and the definition of cycle, it is clear that the expression (72.3) is uniquely determined (except for 1-cycles) by the cyclic order of the various cycles and the choice of first elements $a_{i1} \in N$. The *length of the cycle* $(a_{i1}a_{i2}\cdots a_{i\ell_i})$ is ℓ_i, and the arrangement of the cycle lengths in increasing order of magnitude $\ell_1 \leq \ell_2 \leq \cdots \leq \ell_m$ is the *type of the permutation* α. For example, $(12)(34) \in S_4$ has type (2,2). When two permutations have the same type, we also say that they have the *same cycle structure*. The concept of cycle structure yields a nice characterization of conjugate permutations.

72.4 Theorem

Let $\alpha, \beta \in S_n$. Then α and β are conjugate if and only if α and β have the same cycle structure.

In addition to the disjoint cycle factorization (72.3), we may also factor any permutation into *transpositions* (2-cycles). This can be seen by factoring each cycle in (72.3):

$$(a_{i1}a_{i2}\cdots a_{i\ell_i}) = (a_{i1}a_{i2}) \circ (a_{1i}a_{i3}) \circ \cdots \circ (a_{i1}a_{i\ell_i}).$$

While a permutation α has many factorizations into transpositions, the parity (odd or even) of the number of transpositions is independent of the factorization. This fact yields a partition of S_n into two sets, one containing the even permutations and the other the odd permutations — $\alpha \in S_n$ is an *even permutation* if $\alpha = t_1 \circ \cdots \circ t_{2k}$ where each t_i is a transposition, and otherwise, α is an *odd permutation*.

72.5 Theorem

For $n \geq 2$, the symmetric group S_n is generated by the set of transpositions. The subset of permutations that may be factored into a product of an even number of transpositions is a normal subgroup A_n of index 2.

We call the normal subgroup A_n of S_n the *alternating group* of degree n. That the order $|A_n|$ of A_n is $n!/2$ follows from 72.5.

72.6 Theorem

(*i*) *For $n \geq 3$, the alternating group A_n is generated by the set of 3-cycles.*
(*ii*) *If $n \geq 5$, then A_n is generated by the permutations of type $(2,2)$.*
(*iii*) *For $n = 4$, the set of permutations of type $(2,2)$ generates a normal subgroup (of S_4) of order 4.*

§73 Centralizers of Permutations as Direct Products

(To give context to the centralizer theory developed in Chapter 4, an understanding of the theorems and constructions in this and the following section is a must.)

The *centralizer of a permutation* $\alpha \in S_n$ is the subgroup of S_n consisting of those permutations that commute with α. For understanding which permutations commute with α, we have the following theorem, which states that $\beta \in S_n$ commutes with $\alpha \in S_n$ when β permutes the cycles of α while preserving their (α-induced) cyclic ordering.

73.1 Theorem

Let $\alpha, \beta \in S_n$, and let

$$(a_{11}a_{12}\cdots a_{1\ell_1}) \circ (a_{21}a_{22}\cdots a_{2\ell_2}) \circ \cdots \circ (a_{m1}a_{m2}\cdots a_{m\ell_m})$$

be the cycle decomposition of α. Then $\alpha \circ \beta = \beta \circ \alpha$ if and only if α has cycle decomposition

$$(b_{11}b_{12}\cdots b_{1\ell_1}) \circ (b_{21}b_{22}\cdots b_{2\ell_2}) \circ \cdots \circ (b_{m1}b_{m2}\cdots b_{m\ell_m}),$$

where $a_{ij}\beta = b_{ij}$ for all indices i and j.

For $\alpha \in S_n$, let K be the set of positive integers k for which there exists a k-cycle in the cycle decomposition of α; and for each $k \in K$, suppose N_k denotes the set of letters a_{ij} with $\ell_i = k$. Then for $\beta \in S_n$ such that $\alpha \circ \beta = \beta \circ \alpha$, it follows that β leaves N_k invariant, i.e., $N_k\beta = N_k$. In short, each member of $C(\alpha)$ leaves N_k invariant.

73.2 Theorem

Let α be a permutation ($\alpha \in S_n$). A permutation $\beta \in S_n$ commutes with α if and only for each $k \in K$, $N_k\beta = N_k$ and the restrictions β_k and α_k, of β and α to N_k, commute.

Since $N_k \cap N_{k'} = \emptyset$ for $k \neq k'$, we may view $C(\alpha)$ as a direct product

$$C(\alpha) = \times_K C_k(\alpha_k)$$

where $C_k(\alpha_k)$ is the centralizer (relative to S_{N_k}) of $\alpha_k \in S_{N_k}$. Since each α_k is regular (all cycles in the cycle decomposition of α_k have the same length), $C(\alpha)$ is a direct product of centralizers of regular permutations.

§74 Centralizers of Regular Permutations as Wreath Products

By 73.2, the study of $C(\alpha)$ is reduced to a study of $C(\alpha_k)$ where α_k is regular. So we shall assume that α is regular — the cycle structure of α has the following form:

$$(a_{10}a_{11}\cdots a_{1,\ell-1})\circ(a_{20}\cdots a_{2,\ell-1})\circ\cdots\circ(a_{m0}\cdots a_{m,\ell-1})\in S_n.$$

For such an α, we may view $C(\alpha)$ as a wreath product: To motivate this view, note that from 73.1 we may construct a $\beta \in S_n$ that commutes with a given $\alpha \in S_n$ by following a two-step process. First, permute the cycles of α. Second, for corresponding (matched by our first step) cycles of α, choose a bijection β that preserves the α-induced cyclic orderings. Such two-step processes generally yield to some type of wreath product.

To see exactly which wreath product, note that $n = m\ell$, define $P = \{1, 2, \ldots, m\}$, and consider the symmetric group S_m. This choice of S_m allows for permuting the first indices on the letters appearing in the ℓ-cycles of α. For encoding shifts of second indices, select the group $G = Z_\ell$, the cyclic group of order ℓ. Then, with G^P denoting the set of all functions $P \to Z_\ell$, define the *wreath product* Z_ℓ wr S_m as the set $W = G^P \times S_m$ with multiplication $W \times W \to W$ given by

$$(f,t)(g,u) = (fg^t, tu) \quad \text{where } p(fg^t) = pf + (pt)g \in G.$$

Since the binary operation $W \times W \to W$ is associative, W is a semigroup. Furthermore, since W has an identity element $1_W = (c, (1))$, where c is the constant map $P \to G$ whose value is the identity 0 of G, we see that W is a monoid. And since each $(f,t) \in W$ has an inverse (g, t^{-1}), where g may be defined via the equation $(pt)g = -pf$, for $pt \in P$, the monoid W is a group.

74.1 Theorem

Let $\alpha \in S_n$ be a regular permutation whose cycle decomposition contains m cycles, each of length ℓ. Then the centralizer $C(\alpha)$ is isomorphic to the wreath product Z_ℓ wr *S_m.*

In passing, let us reflect on the meaning of $(f,t) \in W$ in the context of 74.1. First, think of $t \in S_m$ as a matching of the ℓ-cycles appearing in the cycle decomposition of α. Second, for each $p \in P = \{1, 2, \ldots, m\}$, think of $pf = s \in G = Z_\ell$ as initializing a cyclic-order-preserving map of the pth ℓ-cycle onto the (pt)th ℓ-cycle. For example, consider the disjoint cycle decomposition of the regular permutation

$$\alpha = (i_0i_1i_2i_3)\circ(j_0j_1j_2j_3)\in S_8.$$

Then, because there are two cycles, $P = \{1, 2\}$, and, because the cycles have length 4, $G = Z_4$. In this case, suppose $t = (1, 2) \in S_2$ and $(1, 3) \in f \in G^P$. Then $(f, t) \in W$ where t matches the first 4-cycle with the second, and $1f = 3$ induces the map $i_0 \mapsto j_3$, $i_1 \mapsto j_0$, $i_2 \mapsto j_1$, and $i_3 \mapsto j_2$.

As a corollary of 74.1, we may calculate the orders of centralizers in S_n.

74.2 Corollary

If the cycle decomposition of $\alpha \in S_n$ contains exactly n_k k-cycles, then $n = n_1 + 2n_2 + \cdots + nn_n$ and the order of $C(\alpha)$ is

$$\Pi_{i=1}^{n}(n_i!)k^{n_i}.$$

§75 Semigroups

Throughout this section, we shall use S to denote a semigroup.

We may always *adjoin an identity* to S by first choosing an element $1 \notin S$ and then defining a multiplication on $S \cup \{1\}$ to be multiplication on S with the additional products

$$1a = a1 = a \quad (a \in S \cup \{1\}).$$

In practice, we may indicate this construction by writing "S^1", which means $S^1 = S \cup \{1\}$ if S has no identity element, and $S^1 = S$ otherwise.

An element $z \in S$ is called a *zero of S* if $za = az = z$ for every $a \in S$. We may also *adjoin a zero* to S by first choosing an element $z \notin S$ and then defining a multiplication on $S \cup \{z\}$ to be multiplication on S with the additional products $za = az = z$ for every $a \in S \cup \{z\}$. In practice, we may indicate this construction by writing "S^z", which means $S^z = S \cup \{z\}$ if S has no zero element, and $S^z = S$ otherwise. And when we prefer "0" instead of z, we use "S^0" instead of S^z. But independent of our choice of notation, any semigroup can have at most one zero.

If S has a zero 0, then it is customary to use S^* to denote the set $S - \{0\}$ together with a partial multiplication $S^* \times S^* \to S^*$ defined only for pairs $(a, b) \in S^* \times S^*$ where $ab \neq 0$. Such a structure is an example of a partial groupoid. On the other hand, if S has a zero and all products are equal to zero, then S is a *null semigroup*.

Suppose $\emptyset \neq A \subset S$. Then the *subsemigroup T of S generated by A* is the intersection of all subsemigroups of S that include A. If $T = S$, then S is said to be *generated by A*, and A is a *set of generators for S*. If $A = \{a\}$ is a singleton set, then the semigroup generated by A is the *cyclic semigroup $\langle a \rangle$ generated by a*. If S is a monoid and $\emptyset \neq A \subset S$, then A *generates S as a monoid* if no proper submonoid of S is a superset of A.

An element $e \in S$ is an *idempotent* if $e^2 = e$. A semigroup is *idempotent* (or a *band*) if all its elements are idempotents. For $a, b \in S$, we say that *a and b commute* if $ab = ba$. If all its elements commute, then S is called a *commutative semigroup*. Since idempotents in an inverse semigroup commute, every inverse semigroup S contains a commutative idempotent semigroup E_S, which consists of all idempotents in S.

For any semigroup S, there is a *natural partial ordering* of the set E_S of idempotents in S, namely,

$$e \leq f \quad \Leftrightarrow \quad e = ef = fe \quad (e, g \in E_S).$$

It is easy to show that this ordering "$\leq$" on E_S is a partial ordering. Moreover, when S is an inverse semigroup, we may view $(E_S, \leq)$ as a special kind of lattice, namely, a *lower semilattice*, which is a partially ordered set E where every two elements $e, f \in E$ have a greatest lower bound $e \wedge f$.

75.1 Proposition

If E is a commutative idempotent semigroup. Then E is a lower semilattice under the partial ordering

$$e \leq f \quad \Leftrightarrow \quad e = ef.$$

Conversely, if E is a lower semilattice, then E is a commutative idempotent semigroup under the operation $ef = e \wedge f$.

The equivalence given in 75.1 provides just cause for calling commutative idempotent semigroups *semilattices*.

Turning to transformation semigroups, we let X be a set and say that a function α mapping a subset D of X into X is a *partial transformation of* X. The set D is the *domain* of α and is denoted $\mathbf{d}\alpha$. The set R comprised of those $y \in X$ such that $y = x\alpha$ for some $x \in D$ is the *range* of α and is denoted $\mathbf{r}\alpha$. The cardinal number of $\mathbf{r}\alpha$ is the *rank of* α. The *empty transformation* is the empty set $\emptyset$ and may be viewed as the partial transformation whose domain $\mathbf{d}\emptyset = \emptyset$ and whose range $\mathbf{r}\emptyset = \emptyset$. The empty transformation $\emptyset$ is a zero in the semigroup of *partial transformations* PT_X *of* X, where PT_X is the set of all partial transformations of X and multiplication is function composition. In the special case where X is finite and $X = \{1, 2, \ldots, n\}$, we denote PT_X as PT_n.

One subsemigroup of PT_n is the *symmetric inverse semigroup* C_n, which consists of all $\alpha \in PT_n$ that are one-one. The members of C_n shall be called *charts* (as in manifold theory) instead of the usual *partial one-one transformations* used in algebraic semigroup theory. As we did above, we shall write the charts on the right of the argument.

Another subsemigroup of PT_n is *the full transformation semigroup* T_n *of* $\{1,\dots,n\}$, which consists of all $\alpha \in PT_n$ that have $\mathbf{d}\alpha = \{1,\dots,n\}$. A similar definition of the full transformation semigroup T_X of (an arbitrary set) X is obvious.

One of the most popular classes of semigroups is the so-called regular semigroups — S is a *regular semigroup* if for every $a \in S$ there is a $b \in S$ such that $aba = a$. A regular semigroup S whose idempotents commute is called an *inverse semigroup*.

§76 Semigroup Morphisms and Congruences

Let S and T be semigroups and suppose $\phi : S \to T$ is a function that *preserves multiplication*, i.e.,

$$(ab)\phi = (a\phi)\,(b\phi) \quad (a, b \in S).$$

Then ϕ is called a *morphism*, a *homomorphism*, or a *representation*. An injective morphism $S \to T$ is a *monomorphism*, a *faithful representation*, or an *imbedding* (of S into T); a surjective morphism is an *epimorphism*; and a bijective morphism is an *isomorphism*.

A bijection of a semigroup S onto a semigroup T is an *antiisomorphism* if ϕ reverses the multiplication, i.e.,

$$(ab)\phi = (b\phi)\,(a\phi) \quad (a, b \in S).$$

If S is also equal to T and if $\phi^2 = 1_S$, where 1_S denotes the identity map $S \to S$, then ϕ is an *involution*.

An equivalence relation ρ on S is a *left* (*right*) *congruence* on S if for $a, b, c \in S$, $a\,\rho\,b$ implies $ca\,\rho\,cb$ (respectively, $ac\,\rho\,bc$). Moreover, ρ is a congruence on S if it is both a left and a right congruence on S. A congruence ρ on S is a *proper congruence* if it is neither the universal congruence ω nor the identity congruence 1_S.

76.1 Proposition

Let ρ be an equivalence relation on a semigroup S. Then ρ is a congruence on S if and only if for every $a, b, c, d \in S$, when $a\,\rho\,b$ and $c\,\rho\,d$, then $ac\,\rho\,bd$.

For each congruence ρ on a semigroup S, the set S/ρ of equivalence classes becomes a semigroup with multiplication

$$[a][b] = [ab] \quad \text{or} \quad (a\rho)(b\rho) = (ab)\rho,$$

which is well-defined because ρ is a congruence.

One particularly useful kind of congruence is a *Rees Congruence*: Recall that an *ideal* I of a semigroup S is a nonempty subset I of S such that, for every $a \in S$, both aI and Ia are subsets of I. If I is an ideal of S, then ρ_I given by $(a,b) \in \rho_I$ if and only if either $a = b$ or both $a, b \in I$ is *the Rees Congruence on* S *relative to the ideal* I. The quotient semigroup S/ρ_I is a *Rees quotient semigroup* and is denoted S/I.

Ideals play an important role in semigroup theory: A semigroup S is *simple* if it has no proper ideals. (An ideal I of S is proper if $I \neq S$.) A semigroup S with zero is 0-*simple* if $S^2 \neq 0$ and S has no proper nonzero ideals. And a 0-simple semigroup is *completely* 0-*simple* if it has a primitive idempotent. (A nonzero idempotent $e \in S$ is a *primitive idempotent* if it is minimal in the partial ordering on E_{S^*}.) A completely 0-simple inverse semigroup is called a *Brandt semigroup*.

We also have ideal extensions: If I is an ideal of a semigroup S, then S is an (*ideal*) *extension of* I *by* S/I. For example, to say that S is an extension of one Brandt semigroup by another means that S contains an ideal I that is a Brandt semigroup and that S/I is also a Brandt semigroup.

§77 Green's Relations

For any semigroup S, we have Green's relations $\mathcal{L}$, $\mathcal{R}$, $\mathcal{D}$, $\mathcal{H}$, and $\mathcal{J}$. These relations were introduced by J. Green [1], and are defined as follows:

$a\mathcal{L}b \Leftrightarrow$ there exist $x, y \in S^1$ such that $xa = b$ and $yb = a$;

$a\mathcal{R}b \Leftrightarrow$ there exist $u, v \in S^1$ such that $au = b$ and $bv = a$;

$a\mathcal{D}b \Leftrightarrow$ there exists $w \in S$ such that $a\mathcal{L}w$ and $w\mathcal{R}b$;

$a\mathcal{H}b \Leftrightarrow$ $a\mathcal{R}b$ and $a\mathcal{L}b$; and

$a\mathcal{J}b \Leftrightarrow$ there exist $x, y, u, v \in S^1$ such that $xay = b$ and $ubv = a$.

These relations allow one to use "egg-box pictures" to visualize (in a somewhat orderly fashion) an arbitrary semigroup. For instance, in Figure 60.1 we see the egg-box pictures of the full transformation semigroups T_1, T_2, and T_3. Looking at the egg-box picture of T_3, in particular, the cell at "location" 12/3 and $\{1, 2\}$ is an $\mathcal{H}$ class. In the picture of T_3, we see 13 $\mathcal{H}$ classes. On the other hand, we may see the $\mathcal{R}$ classes by erasing all (interior) vertical lines, and the $\mathcal{L}$ classes by erasing all (interior) horizontal lines. There are three $\mathcal{D}$ classes, corresponding to the three egg-boxes, and the $\mathcal{J}$ classes are equal to the $\mathcal{D}$ classes since T_3 is finite.

77.1 Green's Lemma I

Let a *and* b *be* $\mathcal{R}$*-equivalent elements in a semigroup* S *and suppose* $s, t \in S^1$ *such that*

$$as = b, \quad bt = a.$$

If L_a and L_b denote, respectively, the $\mathcal{L}$-classes containing a and b, then multiplication $\rho_s : L_a \to L_b$ on the right by s and multiplication $\rho_t : L_b \to L_a$ on the right by t are mutually inverse $\mathcal{R}$-class preserving bijections.

77.2 Green's Lemma II

Let a and b be $\mathcal{L}$-equivalent elements in a semigroup S and suppose $s, t \in S^1$ such that

$$sa = b, \quad tb = a.$$

If R_a and R_b denote, respectively, the $\mathcal{R}$-classes containing a and b, then multiplication $\lambda_s : R_a \to R_b$ on the left by s and multiplication $\lambda_t : R_b \to R_a$ on the left by t are mutually inverse $\mathcal{L}$-class preserving bijections.

§78 Free Semigroups, Free Monoids, and Free Groups

The concepts of free semigroups, free monoids, and free groups are introduced and discussed in §42. In particular, the idea of presentations for semigroups and groups is discussed in the subsection *Congruences and Presentations* of §42.

§79 Categories

In this section, we adopt the standard of writing operators on the left. A *category* $\mathcal{C}$ consists of a class $\mathcal{O}$ of *objects of* $\mathcal{C}$ and a class $\operatorname{Hom}\mathcal{C}$ of morphisms of $\mathcal{C}$ that satisfy the following conditions.

(*i*) For every ordered pair $A, B \in \mathcal{O}$, there is a set $\operatorname{Hom}(A, B)$ of morphisms such that each morphism in Hom belongs to exactly one $\operatorname{Hom}(A, B)$. A morphism $\alpha \in \operatorname{Hom}(A, B)$ is denoted $\alpha : A \to B$.

(*ii*) For every pair of morphisms $\alpha : A \to B$ and $\beta : B \to C$, there is a unique $\gamma : A \to C$, called the *composition* "$\beta \circ \alpha$" of α and β.

(*iii*) If $\alpha : A \to B$, $\beta : B \to C$, and $\gamma : C \to D$, then $(\gamma \circ \beta) \circ \alpha = \gamma \circ (\beta \circ \alpha)$.

(*iv*) For each object A, there is a morphism $1_A : A \to A$, called the *identity morphism*, such that for any $\beta : B \to A$ and any $\gamma : A \to C$, we have $1_A \circ \beta = \beta$ and $\gamma \circ 1_A = \gamma$.

A *functor* from a category $\mathcal{C}_1$ to a category $\mathcal{C}_2$ is a pair of mappings, one from $\mathcal{O}_1$ to $\mathcal{O}_2$, and the other from $\operatorname{Hom}\mathcal{C}_1$ to $\operatorname{Hom}\mathcal{C}_2$. Both of these mappings are denoted as F, and they satisfy the following conditions.

(*i*) $F(1_A) = 1_{F(A)}$ for all $A \in \mathcal{O}_1$,

(*ii*) if $\alpha : A \to B$, then $F(\alpha) : F(A) \to F(B)$, and if in addition $\beta : B \to C$, then $F(\beta \circ \alpha) = F(\beta) \circ F(\alpha)$.

We denote the identity functor on a category $\mathcal{C}$ by $1_{\mathcal{C}}$, and we say that *categories* $\mathcal{C}_1$ *and* $\mathcal{C}_2$ *are isomorphic* if there exists functors $F_1 : \mathcal{C}_1 \to \mathcal{C}_2$ and $F_2 : \mathcal{C}_2 \to \mathcal{C}_1$ such that $F_2 \circ F_1 = 1_{\mathcal{C}_1}$ and $F_1 \circ F_2 = 1_{\mathcal{C}_2}$.

Bibliography

Aĭzenštat, A. Ja.

[1] *Defining relations of finite symmetric semigroups*, Mat. Sb. **45** (1958), 261–280. (Russian)

[2] *Defining relations in symmetric semigroups*, Leningrad. Gos. Ped. Inst. Učen. Zap. **166** (1958), 121–142. (Russian)

Almeida, J.

[1] *On the membership problem for pseudovarieties of commutative semigroups*, Semigroup Forum **42** (1991), 47–51.

Berge, C.

[1] Principles of Combinatorics, Academic Press, New York, 1971.

Birkhoff, G.

[1] Lattice Theory, Amer. Math. Soc. Colloq. Publ., 1940, revised 1948.

Bondy, J. A. and U. S. R. Murty

[1] Graph Theory with Applications, North Holland, New York, 1976.

Bondy, J. A. and R. L. Hemminger

[1] *Graph Reconstruction — A Survey*, J. Graph Theory **1** (1977), 227–268.

Burnside, W.

[1] Theory of Groups of Finite Order, 2nd edition, Cambridge University Press, 1911; Dover Publications, New York, 1955.

[2] *Note on the symmetric group*, Proc. London Math. Soc. **28** (1897), 119–129.

Cauchy, A. L.

[1] *Mémoire sur le nombre des valeurs qu'une fonction peut acquérir, lorsqu'on y permute de toutes les manières possibles les quantités qu'elle renferme*, J. Ecole Polytech. **10** (1815), 1–28.

[2] *Mémoire sur les fonctions qui ne peuvent obtenir que deux valeurs égales et de signes contraires par suite des transpositions opérées entre les variables qu'elles renferment*, J. Ecole Polytech. **10** (1815), 29–112.

Clifford, A. H. and G. B. Preston

[1] The Algebraic Theory of Semigroups, Math. Surveys No. 7, Amer. Math. Soc., Providence, Vol. I (1961).

[2] The Algebraic Theory of Semigroups, Math. Surveys No. 7, Amer. Math. Soc., Providence, Vol. II (1967).

Climescu, A. C.

[1] *Sur les quasicycles*, Bull. École Polytech. Jassy **1** (1946), 5–14.

Dauns, J.

[1] *Centers of semigroup rings and conjugacy classes*, Semigroup Forum **38** (1989), 355–364.

Dénes, J.

[1] *On some properties of commutator subsemigroups*, Publ. Math. Debrecen **15** (1968), 283–285.

[2] *On transformations, transformation-semigroups and graphs*, in Theory of Graphs, Proc. Coll. held at Tihany, Hungary, Editors P. Erdös and G. Katona, Akadémiai Kiadó, Budapest, 1968.

Dickson, L. E.

[1] *On semi-groups and the general isomorphism between infinite groups*, Trans. Amer. Math. Soc. **6** (1905), 205–208.

Ehrenfeucht, A., T. Harju and G. Rozenberg

[1] *Permutable transformation semigroups*, Semigroup Forum **47** (1993), 123–125.

Eilenberg, S.

[1] Automata, Languages, and Machines, Academic Press, New York, Volume A, 1974.

[2] Automata, Languages, and Machines, Academic Press, New York, Volume B, 1976.

Evseev, J. A. and N. E. Podran

[1] *Semigroups of transformations generated by idempotents with given projection characteristics*, Izv. Vysš. Učebn. Zaved. Mat. **12** (103) (1970), 30–36.

[2] *Semigroups of transformations generated by idempotents with given defect*, Izv. Vysš. Učebn. Zaved. Mat. **2** (117) (1972), 44–50.

Freese, R. and J. B. Nation

[1] *Congruence lattices of semilattices*, Pacific J. Math. **49** (1973), 51–58.

Frobenius, G.

[1] *Über endliche Gruppen*, Sitzungsber. Akad. Wiss. Berlin (1895), 163–194.

Garba, G. U.

[1] *Nilpotents in semigroups of partial one-to-one order preserving mappings*, Semigroup Forum **48** (1994), 37–49.

Gauss, C. F.

[1] Disquisitiones Arithmeticae, Apud G. Fleischer, Lipsiae, 1801.

Gluskin, L. M. and B. M. Schein

[1] *The theory of operations as the general theory of groups*, An historical review of Suschketwitch's 1922 Dissertation, Semigroup Forum **4** (1972), 367–371.

Goldstein, R. and J. Teymouri

[1] *Adjan's theorem and conjugacy in semigroups*, Semigroup Forum **47** (1993), 299–304.

Gomes, G. M. S. and J. M. Howie

[1] *On the ranks of certain finite semigroups of transformations*, Math. Proc. Cambridge Philos. Soc. **101** (1987), 395–403.

[2] *Nilpotents in finite symmetric inverse semigroups*, Proc. Edinburgh Math. Soc. **30** (1987), 383–395.

Green, J. A.

[1] *On the structure of semigroups*, Ann. of Math. **54** (1951), 163–172.

Grillet, P. A.

[1] *Commutative semigroup cohomology*, Semigroup Forum **43** (1991), 247–252.

Hall, T. E.

[1] *On the lattice of congruences on a semilattice*, J. Austral. Math. Soc. **12** (1971), 456–460.

Harary, F.

[1] *The number of functional digraphs*, Math. Ann. **138** (1959), 203–210.

[2] Graph Theory, Addison-Wesley, Reading, Massachusetts, 1969.

[3] *On the reconstruction of a graph from a collection of subgraphs*, in Theory of Graphs and Applications, Academic Press, New York, 1964.

Higgins, P. M.

[1] Techniques of Semigroup Theory, Oxford Science Publications, Oxford, 1992.

[2] *Digraphs and the semigroup of all functions on a finite set*, Glasgow Math. J. **30** (1988), 41–57.

Hille, E.

[1] Functional Analysis and Semigroups, Amer. Math. Soc. Colloq. Publ., Vol. 31, Providence, R.I., 1948.

Hille, E. and R. S. Phillips

[1] Functional Analysis and Semigroups, 1957, revision of Hille [1].

Howie, J. M.

[1] An Introduction to Semigroup Theory, Academic Press, London, 1976.

[2] *The subsemigroup generated by the idempotents of a full transformation semigroup*, J. London Math. Soc. **41** (1966), 707–716.

[3] *Products of idempotents in finite full transformation semigroups*, Proc. Roy. Soc. Edinburgh Sect. A **86** (1980), 243–245.

[4] *Products of idempotents in finite full transformation semigroups: some upper bounds*, Proc. Roy. Soc. Edinburgh Sect. A **98** (1984), 25–35.

Howie, J. M. and M. P. O. Marques-Smith
[1] *Inverse semigroups generated by nilpotent transformations*, Proc. Royal Soc. Edinburgh Sect. A **99** (1984), 153–162.
Jänich, K.
[1] Topology, Springer, New York, 1984.
Jordan, C.
[1] Traité des Substitutions, Gauthier-Villars, Paris, 1870.
Kimble, R. J., A. J. Schwenk and P. K. Stockmeyer
[1] *Pseudosimilar vertices in a graph*, J. Graph Theory **5** (1981), 171–181.
Kocay, W. L.
[1] *On Stockmeyer's non-reconstructible tournaments*, J. Graph Theory **9** (1985), 473–476.
Konieczny, J. and S. Lipscomb
[1] *Centralizers of partial transformations*, to appear.
[2] *Path notation for partial transformations*, to appear.
Lallement, G.
[1] Semigroups and Combinatorial Applications, Wiley, New York, 1979.
[2] Letter dated January 9, 1991, unpublished.
Lallement, G. and M. Petrich
[1] *Extensions of a Brandt semigroup by another*, Canad. J. Math. **22** (1970), 974–983.
Levi, I.
[1] *Normal semigroups of one-to-one transformations*, Proc. Edinburgh Math. Soc. **34** (1991), 65–76.
Levi, I. and W. Williams
[1] *Normal semigroups of partial one-to-one transformations* I, in Lattices, Semigroups, and Universal Algebra, Plenum Press, New York, 1990.
[2] *Normal semigroups of partial one-to-one transformations* II, Semigroup Forum **43** (1991), 344–356.
Liber, A. E.
[1] *On symmetric generalized groups*, Mat. Sb. **33** (1953), 531–544 (Russian).
[2] *On the theory of generalized groups*, Dokl. Akad. Nauk SSSR **97** (1954), 25–38 (Russian).
Lipscomb, S. L.
[1] *Cyclic subsemigroups of symmetric inverse semigroups*, Semigroup Forum **34** (1986), 243–248.
[2] *The structure of the centralizer of a permutation*, Semigroup Forum **37** (1988), 301–312.
[3] *The alternating semigroup: congruences and generators*, Semigroup Forum **44** (1992), 96–106.

[4] *Centralizers in symmetric inverse semigroups: Structure and Order*, Semigroup Forum **44** (1992), 347–355.

[5] *Problems and applications of finite inverse semigroups*, in Proc. Internat. Coll. Words, Languages, Comb., Editor M. Ito, World Scientific Publ., Hong Kong, 1992, 337–352.

[6] *Presentations of alternating semigroups*, Semigroup Forum **45** (1992), 249–260.

Lipscomb, S. L. and J. Konieczny

[1] *Classification of S_n-normal semigroups*, Semigroup Forum **51** (1995), 73–86.

[2] *Centralizers of permutations in the partial transformation semigroup*, to appear in Pure Math. Appl.

Liu, C. L.

[1] Introduction to Combinatorial Mathematics, McGraw-Hill, New York, 1968.

Ljapin, E. S.

[1] Semigroups, Amer. Math. Soc., Transl. Math. Monographs, Providence, R. I., 1963.

Lovász, L.

[1] Combinatorial Problems and Exercises, North Holland, Amsterdam, 1979.

Lyndon, R. C. and P. E. Schupp

[1] Combinatorial Group Theory, Springer, Berlin, 1977.

Malcev, A. I.

[1] *Groupoids*, Mat. Sb. (N.S.) **31** (1952), 136–151 (Russian).

Massey, W. S.

[1] Algebraic Topology: An Introduction, Springer, New York, 1967.

McKay, B. D.

[1] *Computer reconstruction of small graphs*, J. Graph Theory **1** (1977), 281–283.

Mills, J. E.

[1] *The inverse semigroup of partial symmetries of a convex polygon*, Semigroup Forum **41** (1990), 127–143.

Moore, E. H.

[1] *Concerning the abstract groups of order $k!$ and $\frac{1}{2}k!$ holohedrically isomorphic with the symmetric and the alternating substitution-groups on k letters*, Proc. London Math. Soc. **28** (1897), 357–366.

[2] *A definition of abstract groups*, Trans. Amer. Math. Soc. **3** (1902), 485–492.

Munn, W. D.

[1] *The characters of the symmetric inverse semigroup*, Math. Proc. Cambridge Philos. Soc. **53** (1957), 13–18.

Munn, W. D. and R. Penrose
[1] *A note on inverse semigroups*, Math. Proc. Cambridge Philos. Soc. **51** (1955), 396–399.

Papert, D.
[1] *Congruence relations in semi-lattices*, J. London Math. Soc. **39** (1964), 723–729.

Parks, A. D.
[1] *The Fermi and Bose congruences for free semigroups on two generators*, J. Math. Physics, **33** (1992), 3649–3652.

Pastijn, F. and M. Petrich
[1] *Congruences on regular semigroups*, Trans. Amer. Math. Soc. **295** (1986), 607–633.

Paulos, J. A.
[1] Beyond Numeracy, Vintage Books, New York, 1991.

Petrich, M.
[1] Inverse Semigroups, Wiley, New York, 1984.
[2] *Congruences on inverse semigroups*, J. Algebra **55** (1978), 231–256.
[3] Introduction to Semigroups, Bell & Howell, Columbus, Ohio, 1973.

Poole, A. R.
[1] *Finite ova*, Amer. J. Math. **59** (1937), 23–32.

Popova, L. M.
[1] *Defining relations of certain semigroups of partial transformations of a finite set*, Uchen. Zap. Leningrad. Gos. Ped. Inst. **218** (1961), 191–212 (Russian).

Preston, G. B.
[1] *Inverse semi-groups*, J. London Math. Soc. **29** (1954), 396–403.
[2] *Representations of inverse semi-groups*, J. London Math. Soc. **29** (1954), 404–411.
[3] *A note on representations of inverse semigroups*, Proc. Amer. Math. Soc. **8** (1957), 1144–1147.

Rees, D.
[1] *On semi-groups*, Math. Proc. Cambridge. Philos. Soc. **36** (1940), 387–400.

Reilly, N. R.
[1] *Free inverse semigroups*, in Algebraic Theory of Semigroups, North Holland, Amsterdam (1976), 479–508 (Proc. Conf. Szeged University 1976).
[2] *Completely regular semigroups as semigroups of partial transformations*, Semigroup Forum **48** (1994), 50–62.

Rotman, J. J.
[1] An Introduction to the Theory of Groups, Allyn and Bacon, Boston, third edition, 1984.

Ruffini, P.

[1] Teoria Generale Delle Equazioni, Nella Stamperia Di S. Tommaso D'Aquino, Bologna, 1799.

Saito, Tatsuhiko

[1] *Products of idempotents in finite full transformation semigroups*, Semigroup Forum **39** (1989), 295–309.

[2] *Products of four idempotents in finite full transformation semigroups*, Semigroup Forum **39** (1989), 179–193.

[3] *Conjugacy in finite symmetric inverse semigroups*, in Proc. Conf. Ordered Structures and Alg. of Computer Languages, World Scientific, Hong Kong, 1991, 177–179.

Scheiblich, H. E.

[1] *Concerning congruences on symmetric inverse semigroups*, Czech. Math. J. **23** (1973), 1–10.

Schein, B. M.

[1] *Techniques of semigroup theory*, Semigroup Forum **44** (1994), 397–402.

[2] *Products of idempotent order-preserving transformations of arbitrary chains*, Semigroup Forum **11** (1975,1976), 297–309.

Séguier, J.-A. de

[1] Eléments de la Théorie des Groupes Abstraits, Gauthier-Villars, Paris, 1904.

Stockmeyer, P. K.

[1] *The falsity of the reconstruction conjecture for tournaments*, J. Graph Theory **1** (1977), 19–25.

[2] *Tilting at windmills or my quest for non-reconstructable graphs*, Congress. Numer. **63** (1988).

[3] *A census of non-reconstructable digraphs I: Six related families*, J. Combin. Theory Ser. B **31** (1981), 232–239.

[4] *A census of non-reconstructable digraphs II: A family of tournaments*, to appear.

[5] *The reconstruction conjecture for tournaments*, Proc. 6th S-E Conf. Comb., Graph Theory, and Computing, Congress. Numer. **14** (1975), 561–566.

[6] *Which reconstruction results are significant?*, in Theory Appl. of Graphs, Fourth Intern. Conf., Wiley, New York, 1981.

Sullivan, R. P.

[1] *Semigroups generated by nilpotent transformations*, J. Algebra **110** (1987), 324–343.

[2] A Study in the Theory of Transformation Semigroups, Ph. D. Thesis, Monash University, 1969.

[3] *Automorphisms of injective transformation semigroups*, Studia Sc. Math. Hung. **15** (1980), 1–4.

Suschkewitsch, A. K.

[1] *Über die endlichen Gruppen ohne das Gesetz der eindeutigen Umkehrbarkeit*, Math. Ann. **99** (1928), 30–50.

[2] The Theory of Generalized Groups, Kharkow, 1937 (Russian).

[3] *Untersuchungen über verallgemeinerte Substitutionen*, Atti Congr. Intern. Matem. Bologna (1928), 147–157.

Suzuki, M.

[1] Group Theory I, Springer, New York, 1982.

[2] Group Theory II, Springer, New York, 1986.

Symons, J. S. V.

[1] *Normal transformation semigroups*, J. Austral. Math. Soc., Ser. A **22** (1976), 385–390.

Thierrin, G.

[1] *Sur une condition nécessaire et suffisante pour qu'un semigroupe soit un groupe*, C. R. Acad. Sci. Paris **232** (1951), 376–378.

von Neumann, J.

[1] *On regular rings*, Proc. Nat. Acad. Sci. U. S. A. **22** (1936), 707–713.

Wagner, V. V.

[1] *Generalized groups*, Doklady Akad. Nauk SSSR **84** (1952), 1119–1122 (Russian).

Yongzhi, Y.

[1] *The reconstruction conjecture is true if all 2-connected graphs are reconstructible*, J. Graph Theory **12** (1988), 237–243.

Zhang, L.

[1] *On the conjugacy problem for one-relator monoids with elements of finite order*, Inter. J. Algebra and Comp. **2** (1992), 209–220.

[2] *Conjugacy in special monoids*, J. Algebra **143** (1991), 487–497.

Index

Selected Titles in This Series

(Continued from the front of this publication)

ISBN 0-8218-0627-0